SpringerBriefs in Business

For further volumes:
http://www.springer.com/series/8860

Antonia Rosa Gurrieri · Marilene Lorizio
Annamaria Stramaglia

Entrepreneurship Networks in Italy

The Role of Agriculture and Services

 Springer

Antonia Rosa Gurrieri
Marilene Lorizio
Annamaria Stramaglia
University of Foggia
Foggia
Italy

ISSN 2191-5482   ISSN 2191-5490 (electronic)
ISBN 978-3-319-03427-0   ISBN 978-3-319-03428-7 (eBook)
DOI 10.1007/978-3-319-03428-7
Springer Cham Heidelberg New York Dordrecht London

Library of Congress Control Number: 2013953887
Published with the contribution of Fondazione Banca del Monte 'Siniscalco Ceci', Foggia, Italy

Springer is part of Springer Science+Business Media (www.springer.com)

Preface

In the light of the recent dynamics of the recession, sparked by the global economic crisis which began in 2007, the implementation of a course of action leading to growth and a recovery of national economies is urgently needed. With this in mind, we aim to show how an exit strategy involving different economic sectors could be applied and may be replicated in various national economies.

In this volume we have focussed on the potential offered not only by the manufacturing sector but also by the agricultural and tertiary sectors. During the crisis these sectors demonstrated a tendential resilience in the Italian economy, and in certain specific segments there has even been a positive trend.

The volume is divided into three distinct sections: the first deals with the topic of entrepreneurship and organizational forms of enterprises with particular reference to entrepreneurship and networks (A. R. Gurrieri), the second and third sections examine the organizational forms of enterprises (networks) with specific reference to the agricultural sector (M. Lorizio) and the tertiary sector (A. Stramaglia).

Contents

Chapter 1
Entrepreneurship and Network

Abstract The aim of this study is to appraise the role of entrepreneurial networking. The opportunity of operating in a net could be, for all the entrepreneurs involved, both the opportunity for gaining and developing at the level of a single productive unit or the way of sharing risk. Moreover, I believe that the entrepreneurial process could benefit from the advantages deriving from the creation and strengthening of the linkages among networking entrepreneurs. This is evident in Italy where networking entrepreneurship favours the familial dimension as it is the most suitable form for emerging and competing. This is particularly true in the agricultural and tourism sectors, traditionally characterized by local, family and social networks.

Keywords Entrepreneurship · Networks · Socio-cultural barriers

1 Entrepreneurship

Is it possible to speak of new entrepreneurs and entrepreneurship in this period of crisis? Is it possible to recognize a new entrepreneur without any kind of information?

The answer is quite simple: entrepreneurship permits open access to all areas of the economic system and for every type of firm.

Structuration theory (Giddens 1991) sustains that entrepreneurship is a social undertaking. In particular Giddens (1991, p. 204) states: "in seeking to come to grips with problems of action and structure, structuration theory offers a conceptual scheme that allows one to understand how actors are at the same time creators of the social systems, yet created by them... It is an attempt to provide the conceptual means of analysing the often delicate and subtle interlacing of reflexively organized action and institutional constraint".

This chapter is written by Antonia Rosa Gurrieri

A. R. Gurrieri et al., *Entrepreneurship Networks in Italy*,
SpringerBriefs in Business, DOI: 10.1007/978-3-319-03428-7_1,
© The Author(s) 2014

Thus, in this theory agents and social systems coevolve. A social system is composed of the resources and rules used on a day-to-day basis by entrepreneurs in their interactions. Moreover, structuration theory tries to explain how agents create social systems while, at the same time, these social systems create actors.

Therefore, structuration theory believes that this interdependence/duality between entrepreneurs and opportunities is, on the one hand, a constraint and, on the other, a positive possibility for entrepreneurs to start an activity and take opportunities.

It is possible to define entrepreneurship as a dynamic process in which new opportunities are fundamental for creating entrepreneurial ventures. A defective circuit exists so that agents act as sources, and these sources of opportunities affect entrepreneurs.

In literature, Shane and Venkataraman's (2000, p. 218) definition of entrepreneurship as a mechanism by which "opportunities to create future goods and services are discovered, evaluated and exploited", is the most accepted.

Some studies on the topic (Shane et al. 2003; Aldrich 2000) outline the willingness of people to play the game of becoming an entrepreneur. The entrepreneurial process is, per se, an evolutionary mechanism in which a person's abilities to take the opportunity of becoming a new agent (entrepreneur) turns into the capacity of identifying and evaluating an opportunity, in pursuing resources and planning how to use this opportunity. Thus the entrepreneurial process is creative because the entrepreneur re-elaborates resources and opportunities, involving human abilities and intangible factors.

Moreover, for many years researchers (Baron 2004; Aldrich and Martinez 2007) have studied both entrepreneurial factors and, more specifically, what type of person could become an entrepreneur. One of the most important and interesting works is that of Nicolaou and Shane (2009) that studies the genetic elements of an entrepreneur. They recognize four genetic factors (Plomin et al. 1990) that affect an individual's possibility of becoming an entrepreneur. These are (1) the individual capacity of recognizing the favourable condition for entrepreneurship in an environmental scenario; (2) because genes chemically influence the brain, familial transmission of these (from an entrepreneurial point of view) is a pre-requisite for becoming an entrepreneur; moreover, (3) these genes determine the character of each person (extroversion) and (4) the ability of accepting stimuli.

I believe that the compliance with this mechanism (the respect of these four conditions) could be a possibility of engaging the entrepreneurial process.

Through the link between entrepreneurship and resource-based theory it is possible to discover that each entrepreneur uses and develops entrepreneurial capabilities to form a rich environment in which entrepreneurship can develop. However, even if the importance of entrepreneurship is well recognized and sustained, there is little information on the elements that affect it.

Therefore what is the link between a firm and entrepreneurship? Since I believe in the strong relation between these two figures, I also believe that, through both the stabilization of the linkages among entrepreneurs and the valorisation of the social factors, it is possible to create active entrepreneurial networking able to furnish stimuli to the local community, so that the kinetic network could be the driving force of local production.

2 Entrepreneurship and the Firm

Is there a well-known link between modern economic theory and the strategic management and economic entrepreneurial one in the literature?

The answer is clear as only entrepreneurship sometimes uses the theory of the firm. The first hint of the entrepreneurship concept in the theory of the firm was the so called "spectre which haunts economic models" (Baumol 1993, p. 17).

However, these two theories are really interrelated because, in the theory of the market, entrepreneurship and the entrepreneur involve the firm, especially referring to organizational theory. Moreover, the entrepreneur is the agent who personifies the market, thus he/she is the firm.

Therefore, everything centres around the figure of the entrepreneur. In this sense, Foss and Klein (2012) identify this personality in different facets:

- An entrepreneur is a manager because usually in the management of small businesses an entrepreneur is also the manager of new firms, family firms and small ones, his entrepreneurship entails routines and friendship linkages. In this respect, the relations between the theory of the firm and entrepreneurship are superimposed, thus "the theory of entrepreneurship in this approach is the theory of how small business owners organize and manage their assets" (Foss and Klein 2012, p. 4). But this vision has a limitation due to of apparently small business: the idea of considering the entrepreneur-manager operatives in a large unit must be excluded.
- Entrepreneurship as creativity. In a Schumpeterian vision the entrepreneur is the innovator. On this intuition, Ekelund and Hebert (1990, p. 569) state that "...people act as entrepreneurs only when they actually carry out new combinations, and lose the character of entrepreneurs as soon as they have built up their business, after which they settle down to running it as other people run their businesses". However, in this idea the link between firm and entrepreneur becomes less intense as the entrepreneur is only the instrument by which new strategies or products/processes are introduced.
- Entrepreneurship as discovery. In Kirzner's (1992) work the idea that the introduction of new products or strategies by the entrepreneur is in itself a revolution clearly emerges, in the sense that the competition acquires much more qualification. Thus, an entrepreneur who has this role is an architect, or rather, one who is able to capture the discrepancy for gaining in profit is the innovator-creator. However, alertness is not sufficient for realizing profit, as resources are needed.

- Entrepreneurship and leadership. One of the well known economists who discusses "mental models of reality" is Casson (2000). This is a part of literature based on psychology and sociology. The entrepreneur who is able to impose his project due to his communicative ability is the leader of the group. In order to pursue his idea, the entrepreneur needs to coordinate the productive factors in the firm.
- Entrepreneurship as judgment. Entrepreneurship could be read as judgmental decision-making (business decision-making) under uncertainty. Judgment requires capital goods, so an individual without capital goods is not an entrepreneur. Thus the link between firm and entrepreneurship is necessary for defining assets and ownership (transaction costs and property rights).

In economic literature the entrepreneurial process is usually understood as one of the most important factors to explain economic development and growth. This is in line with Hisrch and Peters' (1989) argument where entrepreneurship could be defined as the process of creating new valuable thought.

The instrument of this mechanism is the entrepreneur who has intuition and seizes opportunities.

An entrepreneur is a person who, with his/her abilities, conceives a new marketable company and/or recognizes new opportunities. In literature, many works (i.e. Wade 1995; Fuijta and Thisse 2002) demonstrated that the presence of entrepreneurs in a territory is due to agglomeration economies and the characteristics of social environment (Minniti 2005).

Strategic literature underlines that not all the entrepreneurial opportunities and competitive advantages are ready for entrepreneurs. This situation produces a dysfunction and an asymmetry in information. The ability to discover opportunities depends on an entrepreneur's talent. Changing information, in particular at a social level, suggests the integration of some emotional sensations and cultural ties. Usually these ties are mainly responsible for determining both entrepreneurial activity or networking formation. Moreover, the casual acquaintances, or the Granovetterian weak ties, represent the low-frequency exchange of information. This situation causes a non-redundant exchange of knowledge, characterized by the fact that actors do not know each other. However, at the basis of this scenario are sociological aspects, such as trust, goodwill and moral integrity.

These factors are more evident in a limited territory, usually distinguished by a low level of innovativeness and by the presence of many family firms with a strong tradition in the field in which they operate. The necessity of collaborative techniques in this case permits networks to emerge.

3 Network

The great deal of attention given to cooperation between productive units, collaboration between economic agents and market forces, favours the use of and recourse to networking forms.

Porter (1990, p. 78) defines clusters as "... geographic concentrations of interconnected companies and institutions in a particular field". This structure is characterized by players (companies, individuals and institutions, even when localised in different areas) and informal relations (a competitive type of relation, characterised by co-operation based on a collective vision network).

Other definitions of network can be found in Carayannis and Wang (2004) and in Hakansson and Ford (2002). The former consider a network formation of firms localized in the same territory, while the latter define networks as a type of organisation between productive units as an alternative to market and institutional power (administrative). Also Powell and Smith-Doerr (1994) use the term network both in the social (network) and industrial (network) approach.

Literature on networks often underlines that all firms in the group benefit from know-how and knowledge externalities. This is connected with geographical proximity which helps to create a rapid flow of knowledge since all firms are part of an extensive local network.

Moreover, in dynamic agglomeration economies, entrepreneurs involved in a network can participate in a co-localization process due to the rapid and concentrated exchange of ideas on organization, techniques and production.

This results in both the generation and development of actors, in terms of entrepreneurs, and a preference for the smaller dimension. The networking form allows a rapid circulation of local and international knowledge that passes through diversification links that knowledge implies.

The relation between entrepreneurship and network began approximately 30 years ago. The link between these two elements relies on a social event that includes ties between the entrepreneurial process and the economic-social actors.

Hoang and Antoncic (2003) structured the entrepreneurship-network linkage basing their idea on three principles:

- The content of entrepreneurial relationships (network content); one of the most important factors representing a key element of a network for the entrepreneurial process is the easy access to advice and information. In this way entrepreneurs can identify new opportunities and ideas.
- The governance of these relationships (network governance); the related linkages coordinate the networking exchanges of entrepreneurs. The social element at the basis of this second level is trust, including all the legal contracts that derive from this.
- The pattern of relationships (network structure); this is the size, density, heterogeneity and the centrality of a network.

Many elements circulate in a network, having both a structural component such as information, technology and customer organizations, as well as an emotional one such as friendship, signalling content, legitimacy and reputation. By making use of all these elements an entrepreneur could reduce the time required to identify a strategy and reduce risk. Thus, a positive perception deriving from a networking structure (links) could lead to an exchange of resources.

In a network the co-existence of contacts, exchange relationships and the consequently high level of social links, permits the circulation of many sentiments as well as information that can be acquired by the involved entrepreneurs.

I believe that the presence of a network is fundamental to entrepreneurship because firms in a network can expand through inter-organizational linkages. Moreover, structural holes permit small and medium enterprises to gain advantages by occupying a bridging position in the cluster.

In a network, links among units present either localisation or proximity. Usually, the latter includes the geographical concentration and localisation.

The resource-based theory (Barney 1986; Alvarez 2003) acknowledges the necessity of particular assets within a firm. This type of literature highlights the tacit social strategies as a new competitive advantage for firms in the group. The identification of complete, complex, tacit and path dependence capabilities and resources is a necessary condition for specific investments (Gurrieri 2008).

By localising the group, technology and innovation emerge in a local and endogenous character, clearly visible in knowledge spillovers, scope economies, scale economies, and density, typical of the localisation processes. The localisation process is also characterized by proximity (Rallet and Torre 1999).

In literature the importance of proximity is well documented. Rallet (2002) recognizes a geographical and organizational proximity and the interaction between these two categories generates other types. In fact, Kirat and Lung (1999), define three types of proximity: geographical, organisational and institutional, while Boshma (2005) identifies at least five types: geographical, institutional, cognitive, organisational and social.

In an economic interpretation proximity is both organisational and geographic (Torre and Gilly 2000). Organisational proximity bases its origin on similarity and on the exchange (tacit) of behaviour and competencies, but under opportunism and uncertainty. Thus inter-organisational relations (basic knowledge of each entrepreneur) and intra-organisational (entrepreneurial capacity to coordinate) ones are fundamental.

Geographic proximity is a spatial concept and its presence in a specific territory speeds up routines and competencies.

In the literature there are other types of proximity, such as social, institutional and cognitive. Social proximity is well known (Granovetter 1985; Boschma 2005) because there is a strong link between economic and social aspects due to trust between entrepreneurs. On the other hand, cognitive proximity uses absorptive capacity for favouring the exchange of information among agents.

Institutional proximity (Edquist and Jones 1997) influences limiting uncertainty and transaction costs, thus it is the efficiency of a network.

The network model is very dynamic and by studying the relations and ties among productive units, it is possible reconstruct its knowledge structure. Thus, utilizing the triple-helix approach (Etzkowitz and Leyedesdorff 2000) or the neo-institutional one, it is possible to identify the socio-institutional aspects among entrepreneurs or visualize three dimensions of cooperation. These are (1) technical co-operation, which deals with the production modes adopted by the network; (2)

co-ordination of the different organisations, with regard to the view of the network as structural governance; (3) specific technological co-operation, linked to investments in R&D.

Agglomeration is the result of economies of scope and scale however in real life, the importance of the local social environment plays a strategic role. In fact, when social networks exist and operate, agglomeration economies also work. The presence of a (social) network helps the new entrepreneur to reduce ambiguity. Moreover, the decision to become an entrepreneur requires the courage to face a level of uncertainty and the possibility of failure.

Usually, uncertainty is combined with ambiguity (this is different from risk adversion, because the latter is the curvature of a well-defined utility function). In this situation, even if an entrepreneur has all the information, he/she cannot know the real structure of the situation. Thus, rationality is bounded.

Networks often assume the form of interaction and collaboration among productive units, especially among high technology firms. They contain the elements of the nodes that characterise them and that are related to the position, the type of link, the quantity and flow.

Entrepreneurships involved in the network present specific characteristics and the linkages among the productive units have tangible elements (production transactions and production factors) and non-tangible ones (tacit knowledge and information exchange). Thus, in the group there are firms with asymmetric characteristics linked to their specific abilities, relations between firms, activities and resources.

In the co-localisation process that interests the network formation, entrepreneurs promote a rapid and fluent exchange of ideas on different levels, all interested in the behaviour and dynamic network of the firm. Each entrepreneurship in a network has its own type and level of knowledge that, while providing a competitive advantage, is also made available to the group.

Moreover, in literature there is a particular emphasis on the importance of tacit (learning by doing and using, so linked with a particular place) and codifiable knowledge (ubiquitification, Maskell 2001; this is the shared knowledge of the components of a group). In fact, if we consider the knowledge-based view of the firm (Spender 1996; Pinch et al. 2003), each firm is a storage of information and skills, and each entrepreneur involved in a network contributes to the rise in economic gain that is, per se, not mutual.

The networking process of localisation performs some intrinsic characteristics of the group, such as knowledge spillovers and scope economies, and type and quality of the links among entrepreneurs. Moreover, the collaboration process presents typical relations characterized by both intangible and tangible elements. These are productive factors and transactions and the exchange of information which form the net of productive units with different know how.

Dosi and Kogut (1993), and later Boschma (2005), believe that the approach based on the idea that industrial dynamics are greatly influenced by the integration of the technology co-evolution and the internal organisational forms, is the best one for analysing a network. This is because the perspective is based on the global technological level of the group.

Another mechanism that considers the internal elements of the single unit is proposed by Pavitt et al. (1987) who sustain that size, as entrepreneurial ability to react, could be one of the competitive components of the network. This element has more impact if it is sustained by the absorptive capacity (ability of any organisation to acquire, assimilate, adapt and apply new knowledge) of the firms (Zahra and George 2002) that relies on the existing stock of related knowledge. Being part of a network contributes to increasing the absorptive capacity: a firm with a high level of know how could activate a high level of knowledge transfer.

The firms in a network have long lasting types of relations and linkages (ties) built up over time following the purchase by each firm of tangible elements (resources) and intangible ones (trust between agents, organisational ability).

The mechanism that renders the formation of a net possible consists in spill-overs. They start up the process sustained by agglomeration economies. However, at the basis of a network there is knowledge (with specialization) that is able to increase the inter-organizational linkages. Belonging to a network, especially for SMEs, is an opportunity to emerge thanks to the presence of knowledge spillovers. These help SMEs to surpass the limitations of their small size, because knowledge spillovers generate positive externalities through relational capital and physical proximity.

This networking knowledge is usually codified and not incremental because it has a socio-historical basis and it is spread through relations between actors. Moreover, absorptive capacity and social capital favour the spread and circulation of tacit knowledge and the production of involuntary knowledge spillovers. While social capital is a pre-requisite of a firm, absorptive capacity consists in the ability to capture and re-use knowledge and external information. Thus, absorptive capacity depends on investments in R&D that each entrepreneur makes. Therefore, being part of a network contributes to implementing the existing level of absorptive capacity of each unit.

Loasby (2000) believes that the process of knowledge and learning coming from the link between each single firm, and the context in which it operates, forms the cognitive factors of absorptive capacity. Therefore, the higher the level of each single unit, the higher the transfer of new knowledge among the firms of the group.

Networks of SMEs are usually able to compete at local and international level thanks to the division of labour among firms in the group, knowledge sharing, geographical proximity and agglomeration economies. These elements, in fact, influence the amount and typology of communication and transaction costs.

Moreover, the relevance of social capital is well recognized. It can be defined as a combination of social elements such as rules, networks and institutions that, according to Nahapiet and Ghoshal (1998), favours the efficiency of common and adaptive action of agglomerations. Tura and Harmaakorpi (2005) identify three configurations of social capital:

- structural, which refers to the general elements of a network;
- relational, that identifies only the inter-personal relationships of the net;
- cognitive, that refers to subjective interpretations.

In the literature, social capital is connected with voluntary cooperation and trust. Or rather, an overlapping (Knack and Keefer 1997) of social capital and trust could exist when trust is a measure of the average number of the groups quoted for each region. Moreover, Putnam (2000) suggests two different ways to identify and measure social capital: (1) bonding social capital, that measures only the linkages between homogenous groups; (2) bridging social capital, in which the links between the different groups border on being part of a social network.

Mayer et al. (1995, p. 712) define trust as "the willingness of a party to be vulnerable to the action of another party based on the expectation that the other will perform a particular action important to the trust or, irrespective of the ability to monitor or control that other party". This is particularly true in the context of entrepreneurial network development in which an entrepreneur's ideas are not protected and there are high information asymmetries and adverse selection. Trust in an entrepreneurial network often evolves in the interactions of the actors. The affective trust is the emotional side of trust and consists in the affective aspect of relations, while cognitive trust results when an actor is aware of trusting in the best level of knowledge he/she has.

Therefore, I do not think that it is possible to separate the relationship 'firm-entrepreneur-network' because all are strictly connected and the element they have in common is entrepreneurship, in a wider interpretation (technique and social features).

4 Entrepreneurship and Network

In the literature network formation initially passes from dyadic exchanges and only later do these relations become dense and stable.

The first level of starting exchanges in an entrepreneurial network (Dubini and Aldrich 1991) consists in forming a small circle of contacts and linkages between the new entrepreneur and the involved businesses. At this point, if the result of these linkages is favourable, then the new entrepreneur tries to obtain the necessary resources from his/her family or friends. Obviously, these relations are of a social and emotive nature that relies upon frequent interactions and trust.

Subsequently, these social linkages take the form of a social contract. This situation is the product of the entrepreneurial networking activity that needs to grow, and remaining at an emotive and trust-based level implies a regression. It is at this moment that weak ties emerge, and assume a strategic role due to the fact that they provide non-redundant information.

Finally, the actual formation and growth of a network of entrepreneurs depends on their ability to seek social capital in the new venture. This last phase is the transition (Smith and Lohrke 2008).

Entrepreneurial networks of small and medium enterprises present a high level of flexibility that, through dynamism, produces a higher level of spillovers and capabilities. Some empirical studies (Scharader 1991; Saxenian 1994) show that the geographical and organisational–professional cohesion of firms organised in a cluster, as well as the reputation and technical status of the entrepreneur, are the main factors uniting the firms involved. In this case the informal ties are represented mainly by reciprocity, trust and mutual recognition. In a network, the role of path dependence is fundamental, identifying the location of a possible cluster of entrepreneurs.

The entrepreneurial role in network building is crucial. External relations play a fundamental role in developing the factors that influence the performance of a firm and network formation. The use of external links through inter-firm networks permits the liabilities (of newness and smallness) of entrepreneurial firms to be surpassed (Gurrieri 2013).

Thus, inter-firm networks permit the growth and representation of a model of organization development (Freel 2000). The importance of the relations among firms defines the different kinds of networks. Moreover, the theory of structural embeddedness (Aldrich and Zimmer 1986) using a specific and well defined set of entrepreneurs (network of actors position) and links and a network structure, believes that all could be a constraint or an opportunity.

Another relevant theory in studying networking activity is the structural approach (McEvily and Zaheer 1999) that is based on ties, and in particular on the level (weak, strong) of the ties of the linkages. Thus, network content, or the kind of relations in a net, assumes different aspects in different contexts. Therefore, egocentric networks (links between a single entrepreneur and the others) can mobilize social relations to access external resources. This is not the social network, or the strong ties of a single entrepreneur, but the limit due to the non-correct use of the relations by an entrepreneur.

Social network is important in the start-up phase, helping an entrepreneur to avoid uncertainty and opportunism, especially through predictability and trust. Social networks also help in forming the reputation of networking entrepreneurs. Entrepreneurs, by way of signals, decide to start a specific relation only because this link can help them to gain reputation. On their part, entrepreneurs interested and involved in reputational networks, must offer original and attractive resources for activating this process.

Another type of network among entrepreneurs is the sub-one, which is based on the relational content of the relationships. This is also an organizational network, characterized by the fact that a firm and its founder are inseparable: A firm starts up and grows, and the entrepreneur–founder relations merge.

4.1 *Family Firms*

Recently, Nordqvist et al. (2013), Nordqvist and Melin (2010) and Nordqvist and Zellweger (2010) have published some new research on entrepreneurship in family firms, changing the way of viewing the entrepreneurial process. In fact, in the literature there are many works on the knowledge of family firms, but very few on entrepreneurship.

To date, two different ways of understanding family firms have been recognized:

- the first vision has a positive perspective; it believes that these kinds of unit are characterized by the involvement of many members of the firm that have the ownership and the management (Chirico et al. 2011; Short et al. 2009).
- The second vision (Salvato et al. 2010), the pessimistic one, suggests that the family is an obstacle for the development of the firm. This is because the chief interest of the family in preserving its future, blocks the possibility of taking risks.

The common features of these two interpretations of family firms are that this kind of entrepreneur presents a strong collective identity and a strong emotional involvement. In particular, this latter greatly influences the formation of the family firm's substratum and determines the level of generational involvement, the inclination to change and innovation.

Moreover, entrepreneurship in family firms relies on entrepreneurial orientation (Gurrieri 2008). Chirico et al. (2011) distinguish product innovation, proactiveness and risk taking. Product innovation is the inclination towards the introduction and/or production of new products. Proactiveness consists in the ability to read future needs. Risk taking is the propensity to invest and lose.

Italy is typified by small-scale firms, a characteristic that is always considered a limit to competitive advantage. Moreover, the firm concentration index, that has been decreasing lately, highlights an increasing territorial dispersal of productive units and represents one of the variables involved in the process of globalization. Therefore, if globalization presents cross-border effects, it will not only determine an impulse towards larger scale firms, but also a higher research of dimensional economies of firms, which are preferable to scale economies, in order to raise the concentration.

Speaking about the entrepreneurial figure in SMEs is like speaking about ownership and management contained in one single figure. In this type of unit the entrepreneur-manager has, as a first best choice, the need to provide for himself and his family members, while the second best choice is growth and profit maximization. The idea that entrepreneurs are made and the idea that consequently sustains that everyone can run a small firm, is at the basis of the social development theory. Thus, no one is born an entrepreneur, but everyone could become one. Moreover, the presence of a social network or entrepreneurial culture can only favour a potential entrepreneur and his/her activity. Bygrave (1997) sustains that both internal and external environmental conditions implement the entrepreneurial process. Moreover, because networking activity is crucial for entrepreneurs of SMEs, also rules, convention and membership are important. Thus, in a network, the voluntary mutual shareholding among entrepreneurs is extremely relevant, and determines the strengths of the linkages between firms.

The intensification of trade relations among firms, especially small and medium ones, has captured the attention of experts who are interested in studying the prevailing form of these kinds of relations or networks. These are institutional structures and can efficiently organise all the economic and technological activities that connect and link firms together. An accurate survey of these networks can be useful to build either a flexible production system or a complete organisational form based on the integration of the abilities of each unit.

Although SMEs present problems of dimension and technology, the networks composed of small and medium units seem to sustain competition at local and international level. In particular, in Italy SMEs have to reach a higher dimensional zone in order to become international but, on average, are bigger than the local small dimension. A key role is played by market specialisation, insofar as it is able to offer firms the necessary tools and skills to become international, apart from the turnover or the number of employees. In fact, the fragmentation and segmentation of the market lead to a competitive advantage for the SMEs. Although they are not able to benefit from the large dimension economies of scale, SMEs can cover niches concerning single productive phases and single output categories as well as distinguishing themselves as competitors.

Moreover, the persistent traditional nature and the high demand for typical and traditional products have partly become a competitive advantage for the national industry which seems to find a market in those segments where typicality and quality products are appreciated (Gurrieri and Petruzzellis 2009).

Networks have some position features related to the nodes (type of link and flow). The geographically localised productive activity clusters take advantage of the processes of technological absorption of externalities. Therefore, the positioning of a firm within a group brings the unit to a particular, but not explicit, technological paradigm and to a higher level. For these reasons the productive chain plays an important role, especially for traditional sectors.

In Italy, strong tradition and the historical component play a peculiar role in determining the scale dimension of firms. In fact, Italian entrepreneurs usually prefer to maintain the small or medium level, thus attaining better results.

This is especially true in some sectors, such as agriculture and tourism, usually characterized by family tradition. In these fields, and in those of Southern Italy, it is possible to find all the features of Italian tradition: typicality, familial sharing, historical background and low innovativeness. Both sectors present a networking component that, on the one hand, permits the construction of strong access barriers, while on the other, a network form leads to a closure that improves the basic conditions of the group.

While there is a positive aspect in working in a network, there is also a negative one. In fact, even if collaboration between product units and their entrepreneurs-owners has positive and valid implications, the close interrelation always leads to a lock-in mechanism. The particular aspect of our case is that, even with all these limitations, agriculture and tourism in the South of Italy find a game point in the entrepreneurial networking system. Why is this so? In my opinion, mainly because networking in a low technological sector is not by chance, but is the historical outcome of the sociological factors that influence entrepreneurship. Thus, social and cultural bonds linked with territory are both structural barriers but also winning factors.

References

Aldrich H (2000) Organization evolving. Sage, Beverly Hills

Aldrich HE, Martinez MA (2007) Many are called but few are chosen: an evolutionary perspective for the studying of entrepreneurship. Entrep 25:293–311. doi:10.1007/978-3-540-48543-8_14

Aldrich H, Zimmer C (1986) Entrepreneurship through social networks. In: Sexton D, Smilor R (eds) The art and science of entrepreneurship. Balliner, New York

Alvarez SA (2003) Resources and hierarchies: intersections between entrepreneurship and business strategy. In: Acs ZJ, Audretsch DB (eds) Handbook of entrepreneurship research. Kluwer Academic Publishers, The Netherlands

Barney JB (1986) Strategic factor markets: expectations, luck and business. Strateg Manag Sci 42:1231–1241. doi:10.1287/mnsc.32.10.1231

Baron RA (2004) The cognitive perspective: a valuable tool for answering entrepreneurship's basic. J Bus Ventur 19(2):221–239. doi:10.1016/S0883-9026(03)00008-9

Baumol W (1993) Formal entrepreneurship theory in economics: existence and bounds. J Bus Ventur 8:197–210. doi:10.1016/0883-9026(93)90027-3

Boschma RA (2005) Proximity and innovation: a critical assessment. Reg Stud 39(1):61–74. doi:10.1080/0034340052000320887

Bygrave W (1997) The portable MBA in entrepreneurship. Wiley, New York

Carayannis EG, Wang VWL (2004) Innovation networks and clusters as tech transfer catalysts and accelerators. IAMOT2004, January

Casson M (2000) Entrepreneurship and leadership: studies on firms, markets and networks. Edward Elgar Publishing, Massachussets, USA

Chirico F, Sirmon DG, Sciascia S, Mazzola P (2011) Resource orchestration in family firms: investigating how entrepreneurial orientation, generational involvement, and participative strategy affect performance. Strateg Entrep J 5:307–326. doi:101002/sej.121

Dosi G, Kogut B (1993) National specificities and the context of change: the coevolution of organization and technology. In: Kogut B (ed) Country competitiveness: technology and organizing of work. Oxford University Press, Oxford

Dubini P, Aldrich H (1991) Personal and extended networks are central to the entrepreneurial process. J Bus Ventur 4:11–26. doi:10.1016/0883-9026(91)90021-5

Edquist C, Johnson B (1997) Institutions and organizations in systems of innovation. In: Edquist C (ed) System of innovation. Technologies, institutions and organizations. Pinter, London

Ekelund RB Jr, Hebert RF (1990) A history of economic thought and method, 3rd edn. McGraw Hill, New York

Etzkowitz H, Leyedesdorff L (2000) The dynamics of innovation: from national systems and "Mode 2" to a triple helix of university–industry–government relations. Res Policy 29:109–123. doi:10.1016/S0048-7333(99)00055-4

Foss N, Klein PG (2012) Entrepreneurship and the economic theory of the firm: any gains from trade?. DRUID Working Paper, n. 4

Freel M (2000) External linkages and product innovation in small manufacturing firms. Entrep Reg Dev 12:245–266. doi:10.1080/089856200413482

Fuijta M, Thisse JF (2002) Economics of agglomeration. Cities, industrial location and regional growth. Cambridge University Press, Cambridge

Giddens A (1991) Modernity and self-identify. Polity, Cambridge

Granovetter M (1985) Economic action and social structure: the problem of embeddedness. Am J Sociol 91:481–510. doi:10.1086/228311

Gurrieri AR (2008) Knowledge network dissemination in a family-firm sector. J Socio-Econ 37:2380–2389. doi:10.1016/j.socec.2008.04.005

Gurrieri AR (2013) Networking entrepreneurs. J Socio-Econ, (forthcoming) doi:10.1016/j.socec.2013.09.007

Gurrieri AR, Petruzzellis L. (2009) Knowledge network in unconventional industries. The case of the agri-tourism network. In: Bernhard I. (ed) The geography of innovation and entrepreneurship. Livrena AB, Goteborg, Sweden

Hakansson H, Ford D (2002) How should companies interact in business networks? J Bus Res 55(2):133–139. doi:10.1016/S0148-2963(00)00148-X

Hisrch R, Peters M (1989) Entrepreneurship. Irwin, Homewood, Illinois

Hoang H, Antoncic B (2003) Network-based research in entrepreneurship: a critical review. J Bus Ventur 18:165–187. doi:10.1016/S0883-9026(02)00081-2

Kirat T, Lung Y (1999) Innovation and proximity territories as loci of collective learning processes. Eur Urban Reg Stud 6:27–38. doi:10.1177/096977649900600103

Kirzner I (1992) The meaning of market process. Routledge, London

Knack S, Keefer P (1997) Does social capital have an economic pay-off? a cross country investigation. Q J Econ 112:1251–1288. doi:10.1162/003355300555475

Loasby B (2000) Organisations as interpretative systems. In: DRUID, Conference 15–17 June

Maskell P (2001) Towards a knowledge-based theory of the geographical clusters. Ind Corp Chang 10:921–943. doi:10.1093/icc/10.4.921

Mayer R, Davis J, Schoorman F (1995) An integrative model of organizational trust. Acad Manag Rev 20:709–734. doi:10.2307/258792

McEvily B, Zaheer A (1999) Bridging ties: a source of firm heterogeneity in competitive capabilities. Strateg Manag J 20(12):1133–1156. doi:10.1002/(SICI)1097-0266(199912)20:12<1133:AID-SMJ74>3.0.CO;2-7

Minniti M (2005) Entrepreneurship and network externalities. J Econ Behav Organ 57:1–27. doi:10.1016/j.jebo.2004.10.002

Nahapiet J, Ghoshal S (1998) Social capital, intellectual capital, and the organizational advantage. Acad Manag Rev 23:242–266. doi:10.2307/259373

Nicolaou N, Shane S (2009) Can genetic factors influence the likelihood of engaging in entrepreneurial activity? J Bus Ventur 24:1–22. doi:10.1016/j.jbusvent.2007.11.003

Nordqvist M, Melin L (2010) Entrepreneurial families and family firms. Entrep Reg Dev 22:211–239. doi:10.1080/08985621003726119

Nordqvist M, Zellweger T (2010) Transgenerational entrepreneurship: exploring growth and performance in family firms across generations. Edward Elgar, Cheltenham, UK

Nordqvist M, Wennberg K, Bau M, Hellerstedt K (2013) An entrepreneurial process perspective on succession in family firms. Small Bus Econ 40:1087–1122. doi:10.1007/s11187-012-9466-4

Pavitt K, Robson J, Townsed J (1987) The size distribution of innovating firms in the UK: 1945–1983. J Ind Econ 35:297–316. doi:10.2307/2098636

Pinch S, Henry N, Jenkins M, Tallman S (2003) From industrial districts to knowledge clusters: a model of knowledge dissemination and competitive advantage in industrial agglomeration. J Econ Geogr 3:373–388. doi:dx.doi.org/10.1093/jeg/lbg019

Plomin R, DeFries JC, McCleam GE (1990) Behavioral genetics: a primer. W.H. Freeman, New York

Porter M (1990) The competitive advantage of nations. MacMillan, London

Powell WW, Smith-Doerr L (1994) Networks and economic life. In: Swedberg R (ed) The handbook of economic sociology. Priceton University Press, Priceton

Putnam R (2000) Bowling alone: the collapse and revival of American community. Simon and Schuster, New York

Rallet A (2002) L'Economi de Pro.ximités: Propos d'étape. Etudes Rech 33:11–26. doi:10.3406/ecoru 2004.5470

Rallet A, Torre A (1999) Is geographical proximity necessary in the innovation networks in the era of the global economy? Geo J 49:373–380. doi:10.1023/A:1007140329027

Salvato C, Chirico F, Sharma P (2010) A farewell to the business: championing exit and continuity in entrepreneurial family firms. Entre Reg Dev 22:321–348. doi:10.1080/08985621003726192

Saxenian AL (1994) Regional advantage: culture and competition in silicon valley and Route 128. Harvard University Press, Cambridge

Scharader S (1991) Informal technological transfer between firms: cooperation through information trading. Res Policy 20:153–170. doi:10.1016/0048-7333(91)90077-4

Shane S, Venkataraman S (2000) The promise of entrepreneurship as a field of research. Acad Manag Rev 25(1):217–226. doi:10.2307/259271

Shane S, Locke E, Collins C (2003) Entrepreneurial motivations. Hum Resour Manag Rev 13:257–279. doi:10.1016/S1053-4822(03)00017-2

Short JC, Payne GT, Brigham KH, Lumpkin GT, Broberg JC (2009) Family firms and entrepreneurial orientation in publicly traded firms: a comparative analysis of the S&P 500. Fam Bus Rev 22:9–24. doi:10.1177/0894486508327823

Smith DA, Lohrke FT (2008) Entrepreneurial network development: trusting in the process. J Bus Res 61:315–322. doi:10.1016/j.jbusres.2007.06.018

Spender JC (1996) Making knowledge the basis of a dynamic theory of the firm. Strateg Manag J 17:45–62. doi:10.2307/258190

Torre A, Gilly JP (2000) On the analytical dimension of proximity dynamics. Reg Stud 34(2):169–180. doi:10.1080/00343400050006087

Tura T, Harmaakorpi V (2005) Social capital in building regional innovative capability. Reg Stud 39(8):1111–1125. doi:10.1080/00343400500328255

Wade J (1995) Dynamics of organizational communities and technological bandwagons. Strateg Manag J 16:111–133. doi:10.1002/smj.4250160920

Zahra SA, George G (2002) Absorptive capacity: a review, reconceptualization, and extension. Acad Manag Rev, 27:185–203. doi:dx.doi.org/10.5465/AMR.2002.6587995

Chapter 2
The New Path of Agriculture

Abstract Agriculture has recently evolved, increasing its functions and its role. It has also seen the implementation of multiple innovations. The agricultural sector can primarily implement innovations in its processes. In Italy the most important innovation is the organization of farms into networks. This is principally the result of a different and new entrepreneurship, typical of young Italian farmers. These new organizational processes are the most obvious indicator of the emergence of a "new" paradigm in agriculture, which highlights the social impact and the ethical content of agricultural activity. There is a "new" vision of a "new" socially responsible multifunctional farm, which society also associates with as the provider of public goods. As such, it is protected and financed by the state. In spite of its historical limitations, Italian agriculture more often than not tends to present a united front in its dealings with other sectors. Agriculture has also shown to be more resilient and able to cope better with the current crisis than many other sectors.

Keywords Entrepreneurship · Innovation · Networks · Social and ethical functions

1 Introduction

In recent years Western economic growth has slowed as a result of the economic crisis but, above all, it is no longer considered as an absolute value. Even Italian society has begun to consider a change in lifestyle characterized by the disapproval of waste, sensitivity to environmental sustainability, and awareness of social inequalities. An indicator of this trend is the development of consumption that, in a global negative trend, is characterized by the good performance of sectors like

---

This chapter is written by Marilene Lorizio

A. R. Gurrieri et al., *Entrepreneurship Networks in Italy*,
SpringerBriefs in Business, DOI: 10.1007/978-3-319-03428-7_2,
© The Author(s) 2014

organic food or well-being, while the stage of compulsive consumption—that had characterized the last few years—seems much diminished. In this context, even the relevance of GDP as the exclusive indicator of development has been questioned. The greatest transformation has taken place in the agro-food industry. This sector, driven by a growing criticism over its impact on the environment and on health, and pressured by stringent demands from public opinion and the political world, has increasingly based its own choices on the criteria of social responsibility (Maloni and Brown 2006). Many believe, however, that such entrepreneurial decisions can be imposed by marketing rather than by authentic beliefs, and therefore believe that a stronger role by governments is required to build specific partnerships with private firms to pursue some precise goals. A more critical view of the relationship between the State and the food production system concentrates on the role of consumers as agents of change. Awareness of such a role leads to the concept of the citizen-consumer, a subject who seeks to harmonize individual and social values—intrinsic in the concept of citizenship—with the choices of consumption. In particular the conversion to sustainability may be facilitated by a "pact" between public institutions and civil society, resting on the role of consumer-citizens. In this process, the agricultural system has a very important role. In fact, today agriculture has to resolve new needs that affect the framework in which it works and its institutional conditions. At one time the function of the agricultural sector was almost exclusively to satisfy the demand for foods; nowadays its function is much broader. The agricultural sector today provides considerable non-monetary assets, such as job opportunities and the economic survival of some rural areas with low settlement and rural development. These values are priceless because they relate to the quality of life of present and future generations. Finally, the agricultural sector has the multifunctional role of ensuring compliance with food and environmental safety standards, therefore, it now faces new challenges and difficulties affecting the strategic importance of agricultural planning. Indeed, on the one hand the primary objective of satisfying food demand is faced with new productive philosophies which highlight the variety and the particularity of the inputs, their local feature, their source and their history; on the other, the same primary objective is linked to other strategic objectives, such as environmental and food safety issues. All this involves a different characterization of agricultural activities and of the role of agriculture itself in the various economic and social systems. Farmers are compelled to change their choices in order to realize more and more positive externalities. Therefore, the configuration of the sector is changing to meet new demands of increased environmental and social sustainability which were not covered by the traditional production model. This process of global transition affects historically established organizational paradigms. Its circulation is supported by the market and welfare crisis, by the need to save resources and by the fragmentation of the social structure. The conversion transforms the choices of farmers and public operators, the procedure of creation and distribution of economic value, the organization of common goods, and the functions of the different actors. In Italy, the multi-sectorial nature of farms has grown considerably. Indeed, one way of farming and a single type of farmer no longer exist, they are now many different models of successful agricultural businesses: for example, organic farms or farms that invest in quality products such as Dop/Igp, farms that offer goods and services thus varying their economic activities (farm houses, contractors), or farms that also produce renewable energy (solar, wind, biomass). There were only 233 of these in the year 2000, by 2012 this had increased to 3,485 (Inea 2012 data). Growing consideration of the "social function" of agriculture has increased over the years, not only in Italy but in many European countries. It is seen as a way of life, as heritage, as a cultural identity, as a safeguard of the ecosystem. From this point of view, agriculture produces an added value which is not directly connected to economic features, but rather is linked to issues of inclusion and social cohesion. In recent Italian local systems of social farming, farms are one of the subjects of local networks that are developing and their welfare is a community responsibility for a collection of private and public actors. According to the European Social and Economic Committee this structure can help to respond to a growing demand for acceptance, inclusion and cohesion that is not satisfied by traditional welfare systems, especially in these times of recession. Indeed, social agriculture has shown that welfare can became an income originator and a driving force of development, reversing the conventional timeline in which health has firstly to be produced and then distributed by the State (two-stage welfare). This vision should include the forms of public financial support to the practices of social farming. Given the fiscal crisis of the State, the model of social agriculture appears to be more virtuous. It is, in fact, able to activate local resources, not only financial, such as the propensity of citizens towards more responsible purchasing. Therefore, this new system can ensure sustainable behaviour and choices in new ways. The distinctive factors of the system are communication, research and innovation, attention to demand and lifestyles and, in particular, the creation and coordination of networks between farms. In this respect, associated experiences between farms represent an important indicator of local initiatives and social responsibility, as well as the ability to adapt to market dynamics and deal with institutional changes. These choices are the result of economic strategies carried out by a different and conscious type of entrepreneurship.

The farms react to the modification of the policy targets and of institutional assistance seeking network synergies, which are able to offset some of the institutional imbalances and direct the agricultural system towards new dynamics and a different paradigm.

2   The Evolution of European and Italian Agriculture

The recent financial crisis underlined the difficulties of the global economy. In the world there are constant demographic, climatic and technological changes and globalization connects remote areas which were previously isolated. Agriculture is a victim of these changes while also being responsible for them and for climate change (through production techniques, monocultures and deforestation). The solidity of the global agro food system has been reduced and this has led to a food crisis. The financial crisis emphasized the problem of food security which is a problem of social equality. Health, education and agriculture are the three factors capable of mitigating the problems of poverty and underdevelopment while representing the three fundamental conditions of economic growth. According to the European Commission, in the coming years there will be a sharp increase in global demand for food. This demand has been growing for nearly a decade mostly because of the growing economic development of Asia and Latin America and the WFO estimates forecast an increase of 70 % by the year 2050. This increase in demand leads to a simultaneous increase in supply, particularly from the EU, which is an important player on the international markets and covers about 18 % of world food exports. Therefore, the rural areas are of great importance in the European Union as, on the one hand, they have a more fragile economic performance than urban areas—GDP per capita, employment rates for women, quality of human capital, while on the other hand, they are directly related to environmental protection and to sustainable practices. The European territories present many geographical, economic and social discrepancies and the models of public intervention are very different, but agriculture is no longer synonymous with backwardness. Indeed, some rural regions have evolved through new technologies, progress in transport systems, the increase in tourism, the widespread presence of SME's in the territory and the solidity of urban small systems. As a result, there has been a gradual re-appreciation, also economic, of rural areas. The evolution of rural areas is achieved in three steps: the rural phase, the industrial phase and the post-industrial phase. Each diverges from the others for the role of rural areas, the function of agriculture, the dominant economic sector, and the development of European policies. Before 1970, the only European policy concerned with rural areas was the agricultural one. In the 1970s and 1980s the industrial phase followed: industrial districts were established in rural areas, agriculture became more mechanized and industrialized, but the policy remained segmented and specific. Finally, in the 1990s and from the year 2000 the post-industrial phase expanded: the rural areas and agriculture have become multifunctional and related to various economic activities; the policy also lost its specific nature and has become a territorial development policy. Regulation also reflects this extension of the agricultural firm. The concept of agriculture and, therefore, the structure of the agricultural firm have become more complex and multifaceted. Conventional farming, first identified as crop, livestock and forestry farming, is converted into a number of related activities that strongly characterize the agricultural farm and sector. These "acknowledged" activities are the processing and marketing of home-grown products, provision of non agricultural services through farm resources, promotion of the territory, the rural heritage and forestry and hospitality activities. In this way, the farmer can diversify his activities in response to the multifunctional nature of agriculture established at European level. Lastly, the production of energy from the solar and agro forestry renewable sources has also become part of agriculture. These changes have encouraged the transformation of the sector but have probably over-extended the concept of the agricultural firm. In fact, the acknowledgment of the multifunctional nature of agriculture aimed to increase the skill of farmers so as, to secure additional sources of income to the agricultural firm other than the traditional ones. After the crisis, the European economy continues to evolve and plays a key role within the new development model followed by most countries to overcome the stalemate; a development model disengaged from the old paradigms of mere economic growth and inspired by well- being, social inclusion and sustainability, both in financial and environmental terms. Italian agriculture has transformed and evolved since World War II. Indeed, in the 1960s the Italian agriculture sector was characterized by a dualism in its structure, with small farms opposed to the big capitalistic farms (Bonanno 2012). In the 1980s the emergence of agricultural business destabilized the dualistic structure of the sector (Van der Ploeg 2008). The new and modern agricultural enterprise was still small but, to gain access to innovative technologies aimed at large firms, it was necessary to change both size and organization. This process has led, over time, to a greater uniformity between farms. The evolution of agricultural enterprises and the emergence of new choices based on social cooperation have led to the configuration of local food systems characterized by a multidimensional sustainability, economically, environmentally and socially. This consideration of the economic, social and environmental development of rural areas is associated with the increasing desire of citizens for a better quality of life. In fact, their final purposes are food security, living conditions and health of the territory, leisure and socializing. In this process of ethical growth, the transition course of the agricultural sector has led to innovative tendencies of rural development and to the implementation of agro ecological procedures, in parallel to what happened in the European context. A small scale, multifunctional form of agriculture has emerged. A great knowledge of the organization and diversification of the activities of farms gives a clearer and more accurate perception of Italian agriculture. Indeed, the sector has long been oriented towards more functions, compared to the single primary activity, and therefore it can now produce a wide variety of goods and services. Italian agriculture attempted to respond to the deterioration in farm incomes by diversifying the production processes and activities. In the period of economic and social development, defined as post-productivism, Italian agriculture has also tried to overcome the constraint of productive specialization and standardization and to test new production models based on the segmentation of the product, the supply of service and the improvement of public goods produced by the primary sector (Wilson 2008). In the evolution from productivism to post-productivism the two models coexist, at times intersecting. What emerges is a very composite portrait of Italian agriculture, which is compatible with the processes of development emblematic of the advanced economies and particularly of the European ones. The non-typical activities carried out by the farms can be classified into two categories on the basis of their relations with the primary activity in the strictest sense: the activities aimed at the development of agricultural production— deepening activities—and activities aimed at expanding the set of goods and services offered by farms, beyond the primary activity in the strictest sense— broadening activities (Table 1).

Table 1 Classification of related assets

Deepening	Broadening
Initial processing of agricultural products	Farm holidays
Transformation of plants grown	Entertaining and social activities
Processing of animal products	Educational farms
Woodworking	Crafts
Aquaculture	Production of renewable Energy
Services for livestock	Contract work for agricultural activities
Forestry	Contract work for non-agricultural activities
Feed production	Accommodation of parks and gardens

The broadening activities could often compete with deepening ones over the use of production factors, also because many farms implement the different functions at the same time. They therefore represent composite and multifunctional systems. The existence of small firms is an attribute of the different sectors of the agro-food industry. In this sector, smaller farms almost exclusively practice deepening activities, especially the processing of agricultural products, and when the farm grows, it shifts toward broadening type activities: farming, production and subcontracting services. Usually, diversification is practised mainly by larger farms, because they possess the required entrepreneurial skills and pertinent socio-economic relationships with the territory and institutions (Aguglia et al. 2008). In fact, a fundamental condition for strengthening the processes of diversification is the position of the farms, the network of relationships and the local socio-economic environment. Also financial targeted incentives and more efficient public administration can play an important role in facilitating these processes of diversification. But despite this aspect and the crisis, the Italian agri-food system is the only one to have gone beyond the 2008 levels, demonstrating a virtuous resilience of farms. The agro-food farms, mainly small and medium size, have been less influenced by demand shocks concerning other sectors in previous and more severe years of the crisis, even if there is still a decline in profitability. The most critical factor of the agro-food sector is still the high atomizer which effectively precludes the achievement of economies of scale and the utilization of new production techniques that can reduce unit costs. The small size of a farm is a critical factor especially in the perspective of international competition. In fact, the anti-cyclical nature of demand in the agricultural sector has so far safeguarded farms in times of serious crisis, but in a scenario characterized by a greater weakness of farms the foreign channel becomes decisive to compensate for the difficulties of domestic demand. The amplified size of the business is a pre-requisite for the increase of their impact in foreign markets, together with strategies aimed at consolidating the national retail chains abroad. The low penetration of foreign markets by most of the national farms is certainly linked to some structural limits (reduced size of the business and rigidity of the financial cycle) and to managerial ones, that hamper the propensity to export. A success factor may however be due to the fact that, given the close relations with agricultural production, small farms can offer the values of "Made in Italy", which have great appeal to the consumer. Greater attention to the foreign channel is recorded by some specific agricultural sub-sectors, such as organic farming, which is distinguished because of its performance and its increase in foreign markets. In this sense, the certifications of quality, production efficiency and a high-quality industrial organization—also achieved through cooperation between firms—are essential instruments for realizing good performance, unrelated of the size of the farm. Therefore, to concentrate on these factors is a strategic choice that permits all farms, small, medium and large, to attain positive and lasting results over time. Agribusiness exports increased by 53.1 % from the first quarter of 2007 to the last one of 2011, and represent the greatest performance of the economy as a whole (Istat data). The agro-industrial districts exports have also performed well, with many food districts already being above pre-crisis levels in late 2011. A reduction occurred, however, in the domestic consumption of agro-food products. The decrease in purchasing power and family final consumption are the main difficulties of the economy. In particular, from 2008 families were no longer able to restrain their consumption adequately to counterbalance the consequences of the collapse in income on saving, which has been gradually diminished. In absolute terms, total food consumption increased by more than a third compared to 1970, but a progressive decrease occurred in the weight of food consumption as to total consumption, in accordance with Engel's law. Indeed, according to the 2011 Istat data, this weight, calculated as the percentage ratio between food consumption and total consumption has decreased in the since 1970 when the ratio stood at 35.9–31.6 % in 1980, 23.5 % in 1990, 18.6 % in 2000, recovering slightly in 2010 (19 %). Agriculture over time has been characterized by increasing specialization and intensification of production processes. Furthermore, agricultural policies in support of modernization of the sector have helped the spread of a new business logic. The farmer tries to acquire the most from his resources reallocating production from traditional food to the non- food crops and subsequently using the land for the production of recreational, educational and social services and for the production of solar and wind energy. Between the various development trajectories of the transition process in agriculture a return to farming practices is emerging as is the increasing importance given to the places of production and consumption. Productivity and competitiveness improve in the agricultural sector when the cost of using resources decreases, which is when the production process can use less water, less energy, less fertilizer and pesticides. This practice becomes possible through the adoption of new technologies and managerial innovations that convert the increased productivity into an increase in income for farmers and consolidate their role in the supply chain. A strategic role in this process is played by the so-called managerial farmer, who dedicates at least 30 % of his time to activities not directly connected to the production process but which are related to the organization and management of the firm and to the market. In Italy this type of entrepreneur represents 12 % of farms and 50 % of the production value. The managerial farmer devotes time to knowledge and organizes networks with other farms to upgrade the efficiency process and market access. He also confides in the innovation partners for the growth of his farm and uses modern technologies. Therefore, knowledge and willingness to innovate are the characteristic features of the new agricultural entrepreneurship.

3 Innovation, Knowledge and Entrepreneurship in Agriculture

Knowledge and innovation are significant factors for economic growth. In particular, the adoption of new technologies and the circulation of innovation are two key factors in the development process of all the economic sectors. The argument of innovation has great significance in the agricultural political framework of the European Union. It could be the answer to the rising competitive context originated by the liberalization of markets and to major global challenges resulting from increasing population and scarcity of natural resources. Innovation, especially, is considered a strategic factor capable of revitalizing agriculture and related sectors. It often progresses at a slower rate than expected, as regards both the number of firms that innovate, and the types of innovation. Those predominant in this sector are the "classical" ones, e.g. mechanical, and those regarding the variety of products. However, the innovation related to new markets, direct conversion, new cultivation techniques and certification are less recurrent. The European targets aim to improve the competitiveness of agricultural firms and to increase the supply of food, energy from renewable sources and goods with positive effects on the human health. European initiatives are therefore concentrated on the adoption of new technologies and the circulation of innovation in the agricultural sector. In the second half of the last century global agriculture presented a considerable increase in performances. This increase was caused by investments in research and development and the benefits of this growth have been important for all nations. Indeed, in 50 years (1950–2000) hectare productivity increased by almost 150 %, that of agricultural labour by almost 75 %, and total factor productivity by about 55 % (Istat). This growth is primarily the result of technological progress that has converted the new scientific agricultural knowledge into applied knowledge, i.e. innovations useable in the agricultural process. In this way knowledge becomes a crucial input also in the agricultural sector. Furthermore, the European Commission is trying to reduce the obstacle to the innovative process represented by the distance between search results and the adoption of new practices and technologies by farmers. The intention is to create a system of innovation, based on the correlation of various individuals involved in innovation trajectories of different countries. It seems essential to step up cooperation in networks of agents from different systems and institutional backgrounds. In this system the networks improve their efficiency through the relationship of trust, carry the proper conduct of activities, learning support, divergence regulation and management of intellectual property. In the light of this, innovation processes are not the consequence of a straightforward and planned path, rather they are the outcome of a process of self-organizations of networks. The CAP (Common Agricultural Policy) had a great incentive on the choices of innovation, although their effects are not easily identified. The recent literature has confirmed that the CAP has exercised a considerable impact on investment in innovation in different ways according to the different mechanisms applied (Sckokai and Moro 2009). However it is possible to distinguish some specific areas: training, improving access to networks that promote entrepreneurship and innovation propensity, incentives aimed at reducing costs of innovation or at the lessening of the financial constraints of the enterprise, the selection of operators, strengthening firms with more entrepreneurship, measures on the market conditions (prices, input costs). The Italian agro-food firms have made interesting modernization strategies. The most considerable is the choice to outsource innovation and research and development activities. It is important to understand whether the optimal choice of the company is to produce or acquire innovation. In fact, these options are not inevitably alternative, but complementary, and they depend on the propensity of the firm to innovate. At the theoretical level, the amount of investment in R&D and their specificity play an essential role in establishing the innovation strategies of agro-food systems. Therefore, the greater the specificity of investments, the greater the risk of free riding by potential external partners, which could profit from specific assets of the firm. Consequently, the high risk of specific investments in R&D either determines "in house" solutions or external cooperation relationships characterized by mutual rigorous protection on the use of resources and outcomes. A high degree of concentration in the productive sector of specialization diminishes the propensity to innovate internally, but not externally. The choices of how to innovate also depend on the quality of human capital, on the vitality of the firm and on its aptitude to transmit information and knowledge. In general, agro- food firms implement assorted strategies—i.e. both internal and external innovation— both in the case of process and product innovations. Firms with more years of experience are the ones that have more internal innovation, corroborating the role of accumulation of know-how. An even greater ability, in terms of ICT, amplifies the tendency to innovate internally, and it is not automatically indicative of a greater propensity to outsource innovation. In contrast to what is often theorized, the size of the firm does not influence the choices concerning innovations. One could argue that the choices of agro-food firms in terms of internalization and externalization of innovation are pretty homogeneous. It is necessary, however, to check the capacity of knowledge absorption of firms (absorptive capacity) and their ability to adapt to different socio-economic and institutional contexts (adaptive capacity). Indeed, it is necessary to examine the specific context and the relationship between stimulated transformations and the production system as a whole. The concept of innovation has changed over time: from a "technical novelty produced by science" to elements closely linked to the social, economic and productive system. A systemic vision of innovation which accredits a central function to the subjects and issues rather than the potential innovative contents. From this point of view, knowledge is a collective production and it implies involvement in a practice and not a simple transmission of codified information. The context also plays a key role in developing the skills of the individuals and of the community. Innovation consequently derives from a contextual dimension which embraces the physical, social and productive aspects. It is the result of the contamination between heterogeneous factors, "a process which intersects institutions, producing uncommon and multifaceted relationships between different spheres of activity, which, in turn, are based on the interpersonal relations: the market, law, science and technology" (Callon 1999, p. 84). In particular, the scientific literature has investigated the reasons for the adoption of a specific innovation and of the intensity of innovation in the farms. These include the individual characteristics of the conductor, those of the farm, the prospects of development of the farm and government subsidies. (Ruttan 1996; Janssen and van Ittersum 2007). One of the main individual characteristics of the conductor is the personal attitude towards innovation, which is the timing with which innovations have been adopted in the past. Indeed, this attitude is an indicator of risk propensity, of linkages between human capital and social capital and of the limits of farms, including the ease/difficulty in obtaining credit (Sunding and Zilberman 2001). In particular, in the agricultural sector, product innovations can be achieved through the application of know-how within the firm and are highly specific, while the technological process innovations are less specific and often linked to a transfer to other productive sectors (Capitanio et al. 2009, 2010). At least four groups of characteristics influence the adoption of innovation in agricultural companies (Griliches 1957; Ruttan 1996; Sunding and Zilberman 2001): Individual characteristics of the entrepreneur or his family and the Structural Characteristics of the firm are those "inside" the firm; the Context conditions (market, institutional and cultural context, etc.) are characteristics "outside" the firm. Finally, the Connections with the outside are in some way "crossing" characteristics and they can be located in internal features, although strongly influenced by the context. Qualification and age are the most individual characteristics used to explain the processes of innovation, similarly to those of investment. Indeed, operators with a higher level of education generally have a greater propensity to innovation, while for age there is an inverse relationship. An additional factor much discussed in the literature is the risk propensity (Feder 1980), and other household characteristics, such as number and composition of the family and know-how or aspirations derived from jobs other than farming. The structural characteristics of the firm (firm size, productive specialization, legal form, organization of work, etc.) are linked to the performance and affect the ability to exploit the innovations available and adapt them to the strategic lines of the firm (strategic coherence). Both the financial support and the quality of information are strategic for the circulation of innovation. In general, correct investments in research and technological development have produced technological process innovations which in turn promoted a significant improvement in productivity. In the agricultural context, the dissemination and circulation of these innovations have been encouraged by a number of factors, such as increasing education and information of agricultural workers (knowledge and education) and the services and institutions aimed at informing farmers about the existence of new technological solutions and their correct application (extension of knowledge). It follows the existence of a cause-effect relationship between investment (especially public) on knowledge, research and education with the increase of agricultural productivity. This postulation suggests, in turn, an underlying assumption of a knowledge and innovation system in agriculture. In the traditional European view, this system is based on the research system, on its agents, its institutions and its rules. Therefore, it is characterized by a top down flow of knowledge. Therefore, the innovation is substantially science-based, i.e. a solution "pre-packaged" provided by science in favour of more or less copious applications "downstream", including agriculture. This concept is based on the quality and intensity of technological growth in agriculture in the last century, and indeed, top-down technical solutions have been at the origin of the remarkable productivity growth. This approach has influenced the policy of knowledge in agriculture, concentrated closely on research and on the resolution of its emblematic difficulties. Consequently, there was a greater weight attached to public funding of research, to the role of private research and the viable incentive instruments. There was also great interest in intellectual propriety in knowledge and innovation, to the technological spillover between sectors and areas and their consequences. Therefore the knowledge process in agriculture is based on two paradigms: the top–down model and one founded on network systems. The first is the dominant model for economic strength, size and ability to lobby. It is characterized by established procedures and high predictable funding, and, within it, private research companies require technical changes to the farms. Therefore, the transfer of knowledge and technology shifts from the researchers to farmers through intermediaries; in this dominant model, with a few giants of global dimension, public research and disclosure are often complementary to the private sector, with whom they collaborate. The alternative model has fewer resources and is based on the idea that knowledge is the result of a series of interactions. In practice, research alone fails to produce innovation in the system; innovation is rather the result of the collaborative networks in which the information is exchanged and a learning process occurs. In this model the networks can integrate the production of knowledge with adaptation, counseling and education. This model is made by a myriad of individuals, groups, public bodies and non-governmental organizations, with small funding. However, the results obtained still do not make "system" nor generate a standard modus operandi. Especially in agriculture, innovation is rarely triggered by research, but often it is the response of farmers to the opportunities offered by new and constantly changing markets. The improvements in agricultural productivity not only derive from contributions of science and research, which are then transmitted "downstream" towards productive applications, but they derive also from the process of sharing and propagation of empirical knowledge operating downstream. Such processes of collective knowledge are only partially related to basic research. They are, in fact, linked to diffusion and technical assistance, the increase in the circulation of information, the broader and extraordinary growth in the average education and training of agricultural workers. The collective dimension offers a fertile terrain for the identification and implementation of innovative ideas based on sharing common visions, know-how and practices. In this framework, innovations arising from knowledge are not transferred but contextualized, i.e. adapted to the specific context, the needs of local producers and of particular territories. The resulting process of identification and implementation of local paths of innovation is therefore based on the potential of the community. Consequently, concentrating the considerations only on scientific and technological research and on a codified innovation is a great limitation. Indeed, the real engine of productivity growth of the last century in agriculture lies in peculiar kinds of knowledge and in informal, implicit, pervasive and gradual innovation processes. All this has resulted in spontaneous development and not just a governed and funded one. Widespread knowledge has therefore had a strategic role, which is collective, not incorporated, and sometimes implicit. This type of knowledge produces much more benefits as it is public, is freely available and can be extended to all territories and sectors of application. Moreover, knowledge in agriculture supports the sustainable growth of productivity and improves the resilience of the system, even in an environment of limited resources and climate changes, without the limitations of the traditional agricultural knowledge system. At the basis of this concept of knowledge in agriculture there is the so-called knowledge system thinking (Röling 1992), that is the belief that the innovative performance of agriculture is not the result of a simple, linear and one-directional process, ranging from the production of knowledge (research) to its application, but the result of systemic interactions between different agents and institutions involved in various ways in the production and circulation of knowledge, and its incorporation into innovative and applicable solutions. This vision is based on the quantity and quality (i.e. intensity) of the interactions of the flows of knowledge, more or less incorporated, and of the information within this system. It consists of a network of dynamic and heterogeneous agents which include organized consumers, pressure groups and opinion movers, in short, a wide range of stakeholders. It also transforms the concept of knowledge and the way in which it is exchanged, communicated and implemented to translate into innovation. In this different system, innovation incorporates both the purely technological aspect such as the social and environmental ones, which involve consumers, citizens, all-rural farming communities, institutions and sectors of processing and marketing. Therefore, any "systemic innovation" involves the interaction of all these subjects, the sharing of information and knowledge, continuous processes and distributed learning. The European Commission itself aims to promote innovation through greater integration between the system of knowledge (universities, research centres) and the world of agricultural enterprises. Indeed, in the "Europe 2012" document, the main concern in the short term is to "overcome the crisis", but the long term object is a "smart, sustainable and inclusive" growth, i.e. a growth characterized by competitiveness due to the knowledge, which is environmentally friendly and able to support employment and social cohesion. In fact, the European view is that knowledge and innovation in agriculture are not produced through a "linear path", in which public research makes a new technology available through its easy transmission. They instead seem the result of a systemic approach, the product of a process of creating network, the effect of interactive learning, an agreement among various agents (Röling 2009) which implies equilibrium between new practices, techniques and different methods of organization (Dormon et al. 2007). The World Bank also describes the national system of agricultural innovation as a network of organizations, firms and individuals with the objective of bringing to market new products, new processes and new forms of organization (World Bank 2006). Institutions and policies, in turn, influence the way in which the agents cooperate, communicate, exchange and develop the knowledge. Therefore common ideas, robust links and information flows between the different agents, both public and private, support and encourage institutional incentives to improve cooperation and harmonize markets, legislative frameworks and policies. In this context human capital has a strategic value (Spielman et al. 2008), as recognized in European and national policies, in which it is considered to be the locomotive of development. They state that the human factor is a pattern of individual resources, tangible and intangible, tacit and codified, based on the knowledge and on the skill to upgrade. Furthermore, the knowledge, skills and competencies acquired by individuals improve their living conditions and, at the same time, support the formation of social and economic welfare. Consequently the networks have acquired increasingly significant functions in the processes of technological and organizational innovation. This has occurred for a number of reasons. First, it is difficult for a single firm, albeit large, to possess all the skills required to innovate, so it is necessary to create relations of cooperative research with external agents (competitors, suppliers, customers etc.). In this way, through the relationships founded, the firm has a specific position within the overall network of relationships between firms and between them and other parties. This position affects its market opportunities and its technological knowledge and is therefore a sign of its innovation and competitive performance. Secondly, the recent scientific and technological research is characterized by two interconnected trends: a rising individual specialization and a more frequent use of teamwork. Thirdly, collective networks are vigorous channels through which scientific and technological knowledge is extended in time and space. Currently, the geographical areas in which the new knowledge was produced tend to exploit the results first. This is due to the limited mobility of skilled labour between firms and to the reduction in their propensity to enter into agreements and collaboration with firms at an increasingly geographic distance. In general agri-food firms have both formal relationships with agricultural producers or distributors as well as other contractual relationships with organizations and research consortia, associations of producers and other suppliers. The greater the propensity of the firm in assuming relationships with others, the greater the incentive to form networks, the greater the transfer of innovation and exchange of know-how and information technology. In this way the uncertainty and transactional costs of outsourcing are cut, while increasing the economies of scale and scope related to the development of knowledge and technologies (Teece 1996).

Innovation is therefore linked to entrepreneurship. Entrepreneurship could be defined as the ability to generate value through the pursuit of an opportunity. Some essential requisites are indispensable: it is necessary to recognize an opportunity, to organize a structure capable of extracting value from the various occasions and it is necessary to control the projects in order to restrict their risk and uncertainty. The concept of entrepreneurship in the European policies is considered as the ability to recognize new opportunities and to react to changing external conditions, rather than the ability to raise revenues and profits. This new and more receptive entrepreneurship requires ad hoc training and widespread information. The link between training, information and entrepreneurship has a key importance in the planning of agricultural policies. Therefore, future policies for innovation must also include the procedures for the circulation of information on the results and on the opportunities generated by the research and technological progress in agriculture. However, agricultural innovation involves new skills and a new concept of entrepreneurship, the potential role of which is important to understand the innovation processes and set policy interventions aimed at encouraging innovation in the agricultural sector. The idea of entrepreneurship is applicable to many of the factors previously considered, both structural and individual. While emphasizing the autonomy of some specific attitudinal characteristics previously considered, it also specifies some behaviours directly linked to innovation more explicitly. A summary of the various elements of entrepreneurship is provided by Ross and Westgren (2009). They define it as the process in which a firm searches, discovers and exploits new revenue opportunities, engaging in commercial or innovative activities. The entrepreneurial skills employed to identify and exploit new revenue opportunities imply: (1) readiness in identifying them (alertness); (2) subjectivity and intelligence; (3) decision-making and speed of inclusion of new markets (decisiveness/speed to market); (4) willingness to take risks (uncertainty bearing); (5) aspiring to higher levels of income. The market generally recompenses the firms that are aware of new opportunities, that are capable of judging those which are closest to their capabilities, and that are quick to adapt and able to support an appropriate level of risk. Another crucial feature of the firm is its efficiency in extracting income from the environment in which it operates. This ability may be linked to several factors (technology, know-how, individual characteristics) and explains the firm's performance in a static framework, therefore it is a key element for its survival and development. In this way, entrepreneurship could be the connecting factor between all the elements that affect the propensity of an enterprise to innovate: generational change, training, individual characteristics and network relationships. The description of entrepreneurial behaviour has an added value in that it presents the structured process (and not only the determinants) that lead to innovation. This specific path towards innovation in the agricultural sector also implies processes of diversification and multifunctionality. Nowadays the agricultural sector is faced with a paradigmatic shift based on new conditions: (1) reorganization in terms of its multipurpose function; (2) reformulation of the priorities related to the different functions attributed to the sector (3) identification of new and different trajectories of technical development and technological progress that will characterize the new paradigm. Issues such as the multifunctionality of agriculture and the integrated planning of rural development have enormous implications. In particular the multi-functionality of agriculture involves the ability to respond to the needs of the community, not only in terms of production but also environmentally and socially, and innovation plays a relevant role in this. In addition to food, the agriculture of the future will be able to produce, other non-food goods and services of collective interest, environmental services related to sustainability, landscape and aesthetic values, cultural and leisure services, physical and intellectual welfare. Agriculture, therefore, must be able to ensure food safety and quality, security, well-being, nutritional, environmental and ethical food, and assure their origin and provenance. Sustainability and multi-functionality, however, require the production of knowledge and innovations of a different nature. These are product innovations, (or functions) rather than process ones, organizational and marketing innovations as well as technological ones, more complex innovations and, above all, a wider knowledge than that relating "only" to production processes and "only" to agricultural markets. The knowledge system in agriculture should therefore be re-planned in order to face these challenges and take the opportunities offered by technological change. This re-organization is supported by the affirmation of the new technological paradigm, and of new technological trajectories arising from it. Indeed, the new technological paradigm includes an innovation function (or functional), in addition to product and process innovations. In this way the boundaries of the agricultural sector become less defined and overlap and merge with other sectors, such as the food sector, retrieval, protection and environmental restoration of the sector, the energy sector, the tourism, culture, education and leisure sectors. The main outcome of all these changes in the technological paradigm is precisely the redefinition of the sectors' domains and, therefore, this convergence of sectors that were previously seen as distinct areas. Nowadays this combination between different sectors is identified as a bio-based economy or bio-economy.

4 Italy

In Italy the agri-food industry is the second largest in the country. A total of nearly 3.9 million workers are employed in agriculture, 68 %of which are from the same household (Istat). Italian agriculture is therefore based on a familiar pattern. It has several economic and non-economic objectives, and benefits from different skills depending on the specific production, the type of business and socio-economic context. Today Italy produces, on average, approximately 30 % of its annual food requirements, while the rest is imported. Historically, compared to the overall economy, the agricultural sector has decreased its contribution to GDP, in terms of both production and employment. This is due to some features of the process of development and to structural characteristics of the industry, such as Engel's Law and the lower accumulation capacity and investment in the agricultural sector compared to industry and services. These secular trends of Italian agriculture, rather than being a sign of its progressive marginal nature compared to the rest of

Table 2 Contributions of agriculture production and employment (% of national total)

Years	Output	Added value	Employment
1970	7.6	8.8	18.4
1980	5.2	6.0	14.2
1990	2.9	3.5	8.9
2000	2.1	2.8	4.7
2010[a]	2.9	1.9	4.0

Source Based on Istat data 2011

[a] The respective figures for 2009 were as follows: output 1.7; added value1.8; employment 3.8

the economic system, could be the indicator of a gradual increase in agricultural productivity (Table 2).

However, the Italian position is peculiar in that, while importing the greater part of agricultural commodities—especially grains—it is an important exporter of both processed and preserved products. According to the last Istat national census, there are 1.6 million farms, with over 1.2 million employees (even including those in the fishing industry), 50 billion gross saleable production and 30 billion exports. Over the past few years there has been a significant return to agriculture, the like of which had never been observed in the past. In the first quarter of 2012 nearly 10,000 farms were set up, about 30 % of which are conducted by young farmers under 30 years of age (survey Coldiretti Young Enterprise 2012). In addition, according to the Ministry of Education University and Research, in 2012 the number of students attending agricultural colleges increased significantly (over 11 %), while those attending high schools decreased. Becoming an entrepreneur for a young Italian is almost unachievable because bureaucratic, financial, know-how and technological barriers to entry inexorably reduce the offer. The entrepreneurial "animal spirits" therefore become involved in the founding and closure of micro marginal firms, with low barriers to entry and exit. This choice in agriculture is even more difficult because, in addition to the usual complications, there is the problem of the availability of land. In Italy, land is a restrictive private property. The owners, encouraged by a vicious incentive system that increases the advantage of retaining the property, are not willing to sell, even if their land has been unused or misused for some time. These circumstances damage both agriculture and the country. Moreover, as Adam Smith wrote (1776, p. 71) in "Inquiry into the Nature and Causes of the Wealth of Nations", *agriculture by its very nature does not allow a separation of the work, such as factories, nor allow a separation of one activity from another.* Agriculture is strongly prejudiced by the immobility of the land, and therefore it cannot be outsourced. A farm, especially if family-run, is not relocated because it is intrinsically connected to a specific social context, to a landscape, to a particular local culture. Elsewhere you can make different agriculture, because the earth is different, the social and agrarian landscape is different, the climate, rainfall and wind are different, personal networks are different. For these reasons, the farm is often the priority, the strategic asset for the family. Although Italian agriculture has considerable economic and social relevance, it is still fragile, compressed between the two poles of industry and commerce. There are many difficulties in developing its intrinsic force, partly because of the structural fragmentation of the industry, made up of many small businesses that produce quality but fail to make critical mass. Indeed, the majority of farms (83 %) have single ownerships. This is a sign of the main particularities of the agricultural production system typified by the extensive presence of micro and small enterprises with a productive asset and organizational capital which is often inadequate. The sector's fragmentation also precludes access to economies of scale, impacting on the costs of labour and capital. Finally, the increased variability that characterizes climatic conditions and the increased volatility observed in the markets for agricultural commodities, intensify the peculiarity of the agricultural sector in terms of risk. Therefore, despite its important results, agriculture is an underestimated sector. Its problems are linked to the lack of transparency and fairness in the relations between the various subjects in the supply chain. The ensuing effect is leakage of the economic value at many levels that penalize the weakest part, i.e. the productive one rather than the commercial counterpart. The economic literature is often concerned with the economic effects of the "dependency relationship" of small producers on large enterprises in the large-scale distribution, although it has not yet reached an unambiguous definition of the relationship of dependency. In fact, the asymmetry of size and bargaining power do not always imply the existence of a relationship of dependency and it is often difficult to demonstrate the use of the strong position of large retailers in an anticompetitive way. Generally, the allocation of value among the various agents in the Italian food chain and its level of efficiency do not only depend on the type of relationship between the different operators but also on their number and some structural features, including:

- The dependence on foreign countries for many types of production (primarily agricultural raw materials);
- An excessive atomization of the productive supply, even if in more recent years a clean break has been registered compared to some trends that had occurred in the past, primarily the acceleration of the process of enlargement of the average farm size;
- Consumption patterns of the population (e.g. preference for products with a higher content of service);
- The type of food consumed;
- A concentration in the distribution/sales not yet in line with the main European countries;
- Other factors, such as markets served, level of integration and organization of business processes.

Even the demand for innovation of farms seems fragmented and not in line with the offer. However, recently important changes are being produced in the sector. One of the most 'significant' is represented by a new type of farmer who is, on average, younger, more educated and more motivated, open to innovations and an exchange of relations and cooperation with other farms. Probably the re-affirmation of the sector rests mainly on this intrinsic "innovation'".

4.1 Characteristics of the New Italian Farmer: Age

In the demographic structure of Italian agriculture it is interesting to note the increase in the number of young farmers and the specific characteristics related to their age, such as the type of production on their farms and the economic dimension. The generational replacement is one of the major problems of this sector in Italy, even compared with the rest of Europe. Therefore, the need to promote young forces in the agricultural world is increasingly felt. Farmers under the age of 40 years are considered "young" and those over the age of 55 years, are considered "old", in line with both the specific literature, especially on public interventions (Inea-Oiga 2005, 2009), and with the principal measures of rural development focused on the settlement of young farmers, which poses 40 years as the maximum limit for access to support. According to the impact study published by the EC (European Commission 2011), there are 1.8 million young farmers, 14 % in the 27 EU member states, and they also hold 20 % of the potentially eligible land. According to Eurostat and Istat data, within the European Union, Italy is one of the countries with the lowest number of young farmers: in 2010, only about 5 % were less than 35 years of age, compared, for example, to values higher than 10 % in different European Member States. More specifically, there are about 152.000 farms run by farmers younger than 40 years of age, representing 10 % of the total. Almost half of young farmers are over 35 years, while those under the age of 25 are less than 10,000 units, approximately 6 % of the young and less than 1 % of the total number of Italian farmers. However the presence of farmers over 55 years is more significant reaching almost 60 %. In particular, the percentage of older farmers, over 65 years, nearly 38 % of the total, appears relevant. Even more significant is the figure for the relationship between young people under the age of 35 and farmers over 55 years, because this index reveals what will happen in the near future. In 2010, for Italy the value of this ratio is 0.08. This means that there is only one young farmer for every 10 over 55 years. It is clear that in a fairly short time (10–15 years) a substantial proportion of older farmers will stop working and the question of how many will have a familiar successor will arise. In the family labour force, only 16 % are younger than 40 years old, while nearly 30 % are aged between 40 and 54 years. The data show a small difference in the geographical distribution, with a greater presence of young people in certain geographical areas. The problematic nature of an "ageing" sector is therefore a characteristic feature of Italian agriculture. The ageing of farmers in Italy presents a structural problem and highlights the difficulties of older farmers who want to leave the sector and that of younger people who want to to enter it (Inea-Oiga 2005). The high age of the members of the family occupied on the farm, however, could be a positive indicator of the sector's capacity to preserve employment, at least for family members. However, the ageing of the conductor raises many questions for the future of the sector, even for those least apt to accept the changes. Nevertheless, it seems that a generational change in agriculture is slowly happening. However, there is a significant presence of a range of people aged between 40 and 54 years (approximately 32 %), who are competent in farm management and who can direct the transition to the next generation. Moreover, there are several entrepreneurs driving the farm, who tend to be, on average, younger (mean age 40 years), backed by studies in the economic field, and their entrepreneurial tendency towards farming is the effect of deliberate choices rather than need or inheritance. This aspect shows the new vivacity of the sector, which is in the lowest positions in an ideal European ranking, together with Portugal, even if the difficulty is rather common in the EU countries. Nevertheless, national farms where the holder is young and has a higher level of education have a number of advantages. They are able to invoice 75 % more than the national average in one year, they have an area of more than 54 % higher than average, they often directly manage the processes of transformation and have a greater propensity to organic farming and activation of related tertiary activities (especially agro-tourism activities, aimed at environmental protection and enhancement). The young Italian farmer is principally male (72 %) and the number of women is in line with that of the entire agricultural sector (Istat data). Most of the young conductors are characterized by a high human capital: indeed, more than 10 % of them have a college degree and almost 47 % have a secondary school diploma, while there are virtually no young people without qualifications. The ageing of agriculture is one of the major challenges in which agricultural policy measures should intervene. In fact, the presence of young farmers affects the current production capacity, more so for the future, as well as the sustainability and quality of agriculture. The measures of rural development should focus on the entry of young people in agriculture, but above all for their permanence (Inea-Oiga 2005). The number of local farmers (and of young farmers under the age of 30 years) with a high education, may help not only the transmission of principles, techniques, tools and know-how, but also the production of a learning environment aimed at activating new processes of knowledge creation. This is because a young man is intrinsically more inclined to work in new ways, to experiment with new approaches, skills, and markets. He is less conditioned by the past, is more educated and embedded in the contemporary world and in its many prospects for change. The generational replacement could help farms overcome the crisis, stimulating changes and choices, thanks to a greater sensitivity to environmental and social issues and to greater knowledge.

4.2 Education

The value of education is appreciated, not only at the theoretical level, as being decisive in improving competitiveness and sustainable economic growth. Human capital is in fact a strategic instrument for the efficiency of enterprises and territories. The measures aimed at developing and consolidating human potential are in accordance with the needs of the rural world and with the European policies linked to the "Lisbon Strategy" and the "Europe 2020" for greater European competitiveness and better and more job opportunities. Generally, the financial weight given to measures for the enhancement of human capital in rural development strategies of the Italian regions can be used as a proxy of the importance attributed to these themes. In particular, measures aimed at the development of human potential have been implemented in Puglia, Calabria and Sicily, as part of the 2007–2013 planning. But in the whole national territory there are hopeful signs, especially concerning the figure of the "new" farmer. Currently, the most recurrent educational level among farm managers is the primary school (39 %) followed by middle school (32 %). The proportion of those who have a high school diploma (17.8 %) or a diploma of professional qualification (4.5 %) is not significant; finally, little more than 6 % are graduates. The situation has, however, evolved over time, because in the year 2000 the percentage of conductors with a primary school leaving certificate was 57.5 %, middle school was 23.7 %, while 15.6 % were high school graduates and only 3.4 % of the total number of conductors were university graduates (Istat data 2010). There is, therefore, a shift from lower levels of education to the average levels, and a considerable increase in graduates. This is associated to the outflow of less educated farm managers and to the entry of new generations with higher levels of schooling. It is interesting to note that on organic farms the level of education of farm managers is higher: just over 17 % have, at most, primary schooling, almost 29 % went to junior school, nearly 17 % to high school and more than 32 % are university graduates. However, compared to other manufacturing sectors, the delay in agriculture is still significant. The regional divergences are not very significant, while a more detailed analysis of the data shows a positive correlation between the highest level of education and better economic performance. In fact, the higher the degree of the farm manager, the higher the three-year average (2008–2010) of the added value per farm. This value is below the average of the total added value in farms where the manager has a primary school and junior school level, while the increase in the number of farmers with a higher education degree leads to an increase in the average value and to obvious improvements (Table 3).

Even in agriculture, human capital is therefore a strategic element for the progress of farms and for growth in the overall welfare of the sector. However, its potential stock of knowledge, information and skill is relatively poor at the moment and it is difficult—on the basis of the data- to hypothesize an increase of the same, at least in the public education system, since—because of their age—the

Table 3 AV per holding for the qualification of the farm manager (3-year average 2008–2010), €, %

Qualification farm manager	Euros	Difference with average total
Elementary Schools	45.313	−43
Middle schools	67.056	−15.7
Qualification diploma	89.943	13.1
High school	104.627	31.6
Associate degree or post-graduate	132.633	66.8
VA average total	79.513	

Source Rica (farm accountancy data network) 2011

majority of the farmers are excluded from the ordinary school system. An indicator of the presence of human capital on farms is the level of computerization, which is still underused, but which could be a useful tool to connect with skills and knowledge. This tool can allow the farm to create informal networks and formal relations; it can help social interaction and the formation of functional partnerships for the development of firms and territories. Therefore, the age and education of the farmer are critical for change. The young farmer who takes over a family business suffers internal and external influences which control and shape his way of doing business, and they also tend to repress the inter-generational asymmetries. The background of young farmers is very different and affects the way they think and farm. Their entry is almost always disruptive and produces changes in the organization, in the choices of crops, in the way of producing, in the perimeter of the physical and economic activity. The new farmer is usually an originator of trust, he gives new motivation and develops the farm's capacities, even if he favours the cumulative path, the collection and the review of the inheritance of the father, rather than the rejection of the past. These farmers also have new sensibilities that lead to a lower exploitation of the soil, a more rational use of water or to the production of organic foods. Perhaps the most significant change related to new farmers is the use of innovative agricultural techniques and skills designed to follow the development of services in agriculture. The new farmers are characterized by sensitivity and accumulation of complex knowledge and, sometimes, even by new professional skills that help the process of business diversification and the overflow towards activities linked to the agricultural core. Indeed, many strategies implemented by new farmers are aimed at multifunctionality. It implies solutions and initiatives to promote a model of development different from that based on the exploitation of natural resources and workers. For example, the processing of products and the supply of tourism services represent experiments through which farmers try to realize other development models. A multifunctional farm derives its income from a mixed set of activities: farming, farm animal zoos, photovoltaic energy, and tasting days. The farms specialize in some of these activities, combining them with traditional activities. Sometimes the income produced from these new activities exceeds that resulting from agricultural production, which thus becomes a subordinate activity. This expansion of activities is, on the one hand, a good remedy to the risk of monoculture, to the changes in prices and to the crisis in the sector; on the other hand, it is a way to acquire financial liquidity, to meet professional vocations, to develop human relations. Above all, the 'younger age' and the wider culture of the farmer will promote the exploration of new forms of management/organization. There is a greater propensity of the young farmer to build a network, and then to carry out systematic relationships of collaboration and sharing with other farms and there is also an increased focus on innovation.

5 Organizational Processes in Agriculture: The Role of Network

The necessity to introduce and develop specific organizational innovations locally in the agro-food sector, which is still disordered, is the reason for the implementation of a policy focused on networks. The literature on sustainable local food systems in recent years has explored the different organizations and innovations of so-called alternative agro-food networks, that is, structures of production and consumption which are unconventional compared to standardized ones. Their innovation sits in unusual links, new markets and new forms of communication. Within these structures farmers will put new kinds of cooperation and confrontation in place in order to support direct sales of their products and reinforce their independence. The networks of producers may be intensive or extensive. Intensive ones are located in the territory and progress primarily through the intensification of the internal relations between the nodes, while the link with broader or political networks is weak. The extensive networks instead are formed mainly by links with the existing networks, and then reinforced and expanded locally. Both network strategies are based on cooperation and not on competition, such as the economic and social traditional model. Cooperation between firms takes place on three levels: the exchange of inputs and products, trade in services and the exchange of information and knowledge. The cooperation allows a solution to the common problems and tends to expand towards other events and social sectors. In this way the networks not only generate economic flows, but also promote the circulation of information and values. In agriculture, the presence of these alternative social networks is a central factor in the development of sustainable systems. They are characterized by the combination of economic, environmental and social sustainability, and by the strategy to intensify the interconnection of networks. Thus, networks help to build a new marketing system and also a production system. In addition, they address these systems to the market demand, as well as supporting systems of guarantee or certification and social agriculture. It is important to coordinate the various subjects (farms, industrial enterprises, training and research centres) working in a specific agro-food chain to improve the competitiveness of farms, through the creation of stable organizations. In this way the risks associated with price fluctuations are reduced and the system can ensure impartial redistribution of added value at all phases of the supply chain. These aspects are all the more crucial the more intense the disintegration of EU market policy and the volatility of international markets, where transactions on agricultural commodities are often of a financial and speculative nature. A political answer to these problems is to encourage the expansion of territorial aggregations of farms. In these networks, the links between farms are represented by the possibility of reducing production and transaction costs, establishing synergies both in the processes of production and marketing and in the improvement of techniques more congruent with the environment and with the technologies. A network organization can convert the agricultural market from the context of the spot market, characterized by opportunistic behavior related to diverse information, to a context "almost—organization," where transactions are repeated over time. The advantages of this conversion can be condensed as follows: (1) Better quality products to assemble the needs of the various stages of production (the output of a production stage is the input of the next) and reduction of related transaction costs thanks to stable and formalized supply relations; (2) Greater dissemination of information through the supply chain and faster implementation of product and process innovations that involve modifications in the techniques; management of the relationships between firms throughout the network; (3) Entrepreneurial and financial risks are divided within the network. In this way there is a risk reduction, through contractual agreements. A network policy has an indirect impact on the sector and on the local economy. Porter (1990) considers it as a help, a complementary policy which is more effective in the presence of existing networks, focused on development, and of entrepreneurial initiatives. But it is difficult to assess the effectiveness of different policy measures and their contribution to the development of a firms network. In addition, the development of this organization can depend on various factors, many of which are not controlled by the members of the network, and often the forces of development are related to the randomness and to the inventiveness and talent of one of the members. The organizational models and their evolution can be classified in the theoretical framework of neo-institutionalism, which examines how institutional pressures affect the structure of the organizations. Powell and DiMaggio (1991) argue that different organizational structures, in the long run, are characterized by an increasing homogeneity. The organizations put in place a process of institutional isomorphism: they tend to be similar for a variety of external pressures, but do not always achieve greater efficiency. This process of homologation is due to the existence of a common legislative apparatus that influences choices and behavior of organizations, to technical and legal constraints and to the development of the incentive system, which had a profound impact on the choices of farms. Therefore, over time, organizations tend to model themselves on other organizations, following a mimetic path, not tied to considerations of efficiency. The different forms of economic organization seem to be linked to the institutional framework in which they operate. Italy is rich in different spatial contexts and local production techniques and resources, so each area follows a specific path of development. In this way it produces the phenomenon that the economic literature identifies as "a mosaic type of development". Traditionally, in the aggregation process of farmers, Italy has constraints and contradictions, and the effect is still inadequate. It is still frail, not very definite and does not lead to an effective concentration of supply. In addition, this process happens with diverse intensity and rapidity in the country and in some areas it is weaker. Although enlargement of the average size farms is a positive indicator of their technological and economic progress, the organizations of producers are still small, inadequate to lead the markets in a competitive way and acquire the bargaining power necessary to negotiate with subjects upstream and downstream. Both the cooperative tradition and local institutions have played an important function in supporting, or not, the aggregation process of farmers. In fact, the different organizational structures of agricultural producers are most efficient in areas with a solid cooperative tradition and with an extensive network of associations, while they have strong restraints in areas with less developed kinds of economic organization. This implies the inefficiency of the policy to compensate those cultural features, opportunistic behaviour and inadequate business skills that preclude the aggregation of farmers to compete in the market and develop efficient strategies. In Italian agriculture there are more organizational models, suitable to evolve into a network. One model is present mainly in the North East and includes big farms, well-rooted in the territory. They have a "market task" and organize the production, concentrate the supply and place on the market and the production of the members, in a coordinated way. These organizations can also implement a strategy for systems to achieve the objectives of general interest provided for by national and EU policy, and therefore they represent an "engine" of development and a marker for the area in which they work. Another model is created, however, by small organizations, usually located in the southern regions, with weak management ability that can barely represent the privileged agents on the market. Often they are made to take advantage of the Community policy and therefore their main function is to collect public funding and supervise the Community measures rather than operate on the market and start competitive strategies. A similar kind is represented by organizations created to take advantage of the regional policy. The progress of these different models depends on several factors: (1) the territorial factor: the location of a farm in a context where there was already a robust associative tradition has led to its consolidation; (2) the managerial factor which influences the coordination that these organizations, which are also the governance structures, are able to carry out with associated farms; (3) the identity factor, which explains how the culture and the history of farms affects their heritage of resources and know-how. The variety of aggregations over time is structured in forms of proto-district or district, although with different characteristics. The definition of agro-food districts is based on objective criteria such as the number of local units, the number of employees, the specialization index, a productive context characterized by a significant presence of SMEs and by reticular relations between firms, from which a sense of belonging to the local territory derives. Within these categories there are strongly export-oriented districts and districts which work solely on the domestic market. The main economic and financial indicators of district farms since 2010 show a recuperation in the sector, especially in terms of revenues. In all cases they supply high quality products related to the tradition of a specific territory. Often the production of certified quality goods (PDO, PGI, DOCG, Icgt etc.) is one of the strategic choices which allow the agricultural sector to avoid increasing price competition from third countries and to develop the national food production. In this respect, products with a geographical indication and organic products are the most important and pervasive productions in Italian agriculture. These products are related to the needs originated by rising segments of consumers. These, on the one hand, demand local products linked with the territory and with local food traditions, and, on the other, they focus on production processes which are friendly for the environment and for consumer health. Even in these like-districts, some large players influence the performance of the sector. However, the large number of smaller farms that interact with these players and the relevance of the number of workers confirm the hypothesis that these sectors are organized in a district, probably following a peculiar paradigm which would correspond to the "leader enterprise" model, applied to the district. In recent years, new approaches towards the production and consumption of agricultural products have led to new forms of organization, based on a closer cooperation between producers and consumers. Their relationship is not reduced to economic exchange, but includes also the shared objectives of social and environmental sustainability. The process of strategic change is, firstly, a cultural "revolution", based on processes of co-learning and the ability to overcome the technological, institutional and political barriers. This learning has a social and cooperative nature and the subsequent innovations belong to a collective pattern of ideas and projects. As a result, there has been a significant transformation within the dominant agro-food system. The most interesting changes are the projects sponsored by citizen-consumers, focusing on the social, cultural and environmental objectives. These "critical consumers" carry out ethical demands and are very interested in the history and social property of the products. They are characterized by discontent and scepticism of standard production–distribution systems and want to exert more control over their consumption preferences. So, they seek to harmonize their values and choices and to respect the social value of the act of consumption. Both these players, through the dynamics of learning and communication, reshape the socio-technical system and identify new methodologies, rules and managerial models, going beyond the "mere" critical consumption. They have therefore put in place a series of initiatives of social mobilization, on the conversion of lifestyles and on the model of development, constructing a more composite network. These new types of network are named Civic Food Networks (CFN), a definition that stresses their public spirit, their source and their social aspect. These hybrid networks are characterized by a strong interaction between different subjects that share a number of objectives and they are a sign of prospective changes in the mechanism of governance of the agri-food system. The resulting structure involves both market players: farmers belong to a network, although informal, between producers and consumers that support them. This path of social innovation concerns a particular category of producers, defined as critical producer. They have the ability to produce and implement projects, but especially to coordinate many players, to create a future-oriented virtual network. Their specific capability is defined by Van der Ploeg (2006, p. 56) as agency. The critical producers are aware that the current model of economic globalization is based both on consumption choices and on productive ones. They voluntarily cooperate with the critical consumers. The structures of this type are still not very large, but their own niche position encourages the testing of innovative methods of production, distribution and consumption. They therefore are a significant evolution system, as they produce the alternatives in a crisis. Therefore, critical consumers and producers begin with only apparently different needs, get together and work towards the same objective. The expansion of the network, their progressive flexibility and openness to different players with different interests, to new relationships and to related knowledge and procedures are an essential aspect of the innovation process. In these networks, the different players match assumptions and purposes, assemble and divide resources, knowledge, work and relationships, thus leading to further learning processes. The consumer-producer interaction helps the system's response to different needs: the quality of food, empathy, the reciprocal knowledge and organization. They thus complete a process of co-production. The relationship that is established between producers and consumers promotes great transformations, obliges them to acquire new knowledge and skills, to modify the routines, to shape their identity and responsibilities as producers and consumers. In fact, there is a new vision of the production-consumption, which are considered in concert (Rossi and Brunori 2010). The interaction between different players also supports the achievement of sustainability, relating to agricultural procedures and consumption patterns and to food systems as well as to marketing, to the institutional framework and to relations with the territories. Relationships formed in this way promote the development of other forms of ethical consumption and are a factor of the production of other forms of social mobilization on the territory. Other peculiar forms of agricultural organization consist in alternative paths of production/sale, which probably represent the first step of a stable and structured relationship that involves both the supply and demand of agricultural products. These relationships are particular forms of innovation which affect the entire agricultural sector. Alternative Food Network (AFN) is a common term to denote sales systems other than the traditional ones, and until the mid-1990s, much of the literature dealt with alternative channels of sales of agricultural and food products. A report of sale/ purchase directly between producer and consumer is significant when products traded have particular characteristics (Pascucci 2010), so the recognition of their property implies very high transaction costs (organic, fair-trade products, and products from sustainable agriculture). These types of products are classified in the literature as "credence food" (Vetter and Karantininis 2002) and many subjects are involved in their production and distribution. Among the different types of direct sale, the short chain is the most successful initiative; in fact, in a short time it has spread throughout the sector and achieved a socio-technical- economical trajectory, alternative to the traditional one and based on the concentration of the market in large structures. Often the new farmers set up short chain strategies to internalize added value, mainly through organizational types of progressive approach to the final consumer, such as direct sales and vertical integration. Short chain is a model of production and consumption based on the relationship between territoriality, proximity of the goods and consumption, socialization practices, job protection and fair remuneration for those engaged in the food industry, and trusting relationships between producer and consumer. The short chain provides a greater social justice because it modifies the relationship between producers and consumers, bringing them closer to each other, forging more balanced relationships, and permits a more democratic participation of the subjects of the chain (Kirwan 2006; Hinrichs 2003; Tregear 2011). The phenomenon of short supply chains in Italy is developing significantly both for conventional and organic products: from 2007 to 2009 organic farms with direct sales (excluding agro-tourism farms) grew by 31 % (Biobank data 2010). The Istat and Coldiretti data for the same period showed an overall growth of 11 % of the phenomenon, but within a decade or so, i.e. since 2001, direct sales have increased by 57 %. The advantages for the consumer are mainly in the quality assurance of the purchased product. In fact, information asymmetry is compensated by the relationship of trust with the producer, so a formal certification is no longer necessary (Cicatiello and Silvio 2008). This would seem to favour a long-term relationship. In addition, the typical transaction costs of the purchase of quality products are reduced, given the small number of intermediaries (Van der Ploeg 2006; Cicatiello and Silvio 2008). This is also an important benefit for businesses, that are free to trade even in small amounts, because the risks associated with the market power of big sellers decreases (Sini 2010). In addition, these types of sales help the economic development of marginal rural areas and support local small businesses rooted in the territory. In this way they also protect traditional varieties and preserve secular methods transmitted from generation to generation (Battershill and Gilg 1998). Finally, this system can generate new job opportunities for people who do not belong to the agricultural sector. According to the data of the Farm Accountancy Data Network, about 30 % of farms in Italy utilize the short chain channel. It represents a crucial strategy for the realization of entrepreneurial autonomy and this leads farms to try other complementary strategies. This entrepreneurial choice, however, requires specific professional skills and capacity of sale by the farmer, and the need to reorganize the farm, especially with regard to management and logistics linked to direct marketing. The farm, in fact, must not only control the production but also the distribution of the agricultural goods. The solutions implemented have been developing networks between farmers in order to supply a set of different products. Many kinds of short chain have been developed in recent years, each according to trajectories that vary on the basis of economic and also local relations: farmers markets, farm shops, home delivery or macrobiotic restaurants and stores. A distinct experiment is represented by the solidarity purchasing groups, which have a weaker ethical "responsibility" in the purchase, but have greater visibility and are widespread. They are based on continuous interaction between producers and buying groups, which is a focal factor in the improvement of the relationship between supply and demand, and on relations of trust between consumers and producers, built up over time. The farms who supply these buying groups, however, still have difficulty in engaging other farms in order to realize a broad network in the area with which to sustain each other and to better deal with the relationship with consumers. Over time these two alternative forms of organization may interact and support each other. In practice, farmers' markets can be a driving force for the expansion of buying groups and thus further reinforce the relationship between town and country. At the moment, both the sale through buying groups and the sale on the market have required better organization by farmers and also an upgrading of labor resources, proving the power of the most innovative local systems to generate jobs. In strictly economic terms, the sale in alternative short networks allows farmers to keep part of the added value otherwise gained by agents at the different stages further down the food chain. In this way, they counteract the effect of the pressure of revenues on production costs. Another solution to the organizational problem is represented by the parallel consumer associations—called Gas—which coordinate a lot of orders locally. In this way, it is possible to reduce the transport costs and, simultaneously, expand the network. These circumstances have promoted the formation of small networks. Looking to the future, the short chain in Italy will probably increase, thanks to the advantages gained by of farmers and consumers. In addition, generally farmers who start using an alternative sales channel also appear more open to the choice of another innovative sales channel. The process of transition and innovation in agriculture is the result of the attention given to fairness, sustainability, methods of production and marketing. However, it is mainly driven by a "critical production", resulting from the restructuring of farms and of the new farmers, a central figure of these processes of change. The "new critical farmers" have a lot of formal (a medium–high level of education, as a result of education and work experiences) relational (communication skills and team work) and cognitive (autonomy and decision-making) expertise. For them, agriculture—especially organic or biodynamic, but always multifunctional—is an ideological choice, made with sincere conviction. Their most significant innovation is the practice of biodiversity as a strategy to overcome the monocultural model of agriculture, but the new critical farmers also seek to increase the added value through the supply of services. Their goal is not so much an increase in the profit or productivity, but the achievement of autonomy from the market, through the control of the means of production and the times of work and the exploitation of non-commercial circuits for the recruitment and the reproduction of resources. Therefore, the new critical farmers want sustainability instead of productivity, to pursue the variety of crops instead of specialization, they choose to cooperate rather than compete, aim at autonomy rather than efficiency. They choose to work with natural techniques (organic/biodynamic) and work at the defence of biodiversity through the adoption and/or recuperation of traditional types of products and also by testing new crops. Their farms are multifunctional and engaged in direct selling to the consumer, in particular the critical consumer. They also widely use computer technology. Tourism is one of the subsectors in which agricultural innovation is more frequent. In this sector the new farmers construct new organizations and specific forms of collaboration and network. In fact, farmers always react to the demand for tourism by concentrating on their traditional agricultural practice and making it approachable to tourists. In this way they shift from being farmers to being territorial entrepreneurs. In all these cases, the networks between producers help to counteract the increasingly evident phenomenon of the volatility of agricultural prices. According to economic theory, agriculture is a flex price sector, in which the regulation occurs mainly on prices, because the agricultural supply is rigid. Over time, the progressive integration of agriculture with the whole economy, globalization, the concentration of trade in the various stages of the supply chain and the increasing demand have transformed the usual dynamics of agricultural prices. The traditional descending trend in agricultural prices driven by economic growth has been converted in the last few years into a high volatility of those prices worldwide. By having an inadequate transmission mechanism, price changes are not transmitted in full throughout the supply chain because increases/decreases are spread over time, or because the response is not the same. This asymmetrical transmission between agricultural commodity prices and consumer prices is due to the structure of the market: the more oligopolistic the market, the more distortions and impediments there are in the adjustment. Over the years, the phenomenon of the squeeze on agriculture (reduction of revenues and increased costs) has grown, compelling the new/young farmer to seek new strategies and models of production and consumption alternative to those of the agro-industrial dominant system. Innovations to this end have consisted mainly in farm reorganization in multifunctional terms and in atypical marketing methods. The new farmer has thus realized many strategies to achieve economic sustainability, i.e. his autonomy, which represent a strategic factor in the new paradigm of rural development (Van der Ploeg 2008). These strategies are production differentiation, multifunction activities, skill-oriented technologies, the constant increase based on the quantity and quality of labour, and the spread of social wealth. They represent the identifying factors of a model of agriculture aimed at defying the effects of the squeeze resulting from the dominant agro-industrial model and from the technological and regulatory system. In fact, unlike in the dominant model, the new farm is able to set the price and acquire liquidity through direct sales. This ability to differentiate market opportunities and the creation of networks between firms are the identifying factors of a "social structure" different from the dominant economic models. In this system, the greater autonomy that farms are able to attain intensifies the resilience of this new model to produce and sell. The synergy and cooperation that characterize agricultural networks can help to further stabilize the volatility of agricultural prices and rebalance the distribution of added value in the various stages of the supply chain. The new farmers seek to develop new infrastructure, new networks, often informal, with each other and with consumers through which they exchange information and services as well as products. Their social responsibility leads them to operate not only as farmers but also as citizens involved in social programs and associations of environmental safeguarding. In this system, citizens have the privilege and the duty to contribute to the administration of production and consumption (Welsh and MacRae 1998; Hassanein 2003); they therefore establish new relations with public institutions with regard both to the territories—the model of agriculture to support—and to ways through which to intervene—the regulatory system. The result is a creation of an ethical and localized agricultural network as a form of innovation in the agro-food system, capable of supporting the development of social and institutional transformations. Therefore they are important workshop experiences for the progression of the role of civil society. These projects of building a new socio-economic system from the base express a need for different patterns of production and consumption. The various organizational processes in agriculture are based on shared requirements and produce new knowledge and governance patterns and new technical and organizational paradigms

6 Organizational Processes in Agriculture: Social Aspects

The models of innovation in agriculture aspire to modify consumption and to reform social and economic organization into an ethical one. At a micro level, these models also have a strong ethical and social connotation. All the recent events of change around food are analyzed by the theories on social innovation and by the theories of transition (Van der Ploeg et al. 2004; Brunori et al. 2012; Knickel et al. 2009). In them, the role of learning processes that upgrade within the social milieu of the network, in response to common requests or prospects, are strategic, as they are able to activate the cumulative processes of social capital formation. In fact, the social capital that comes from these connections produces an environment favourable to learning and promotes, through the exchange and sharing of experience, the development of knowledge ('peer-to-peer exchange' and 'learning by doing'). These processes, in turn, increase the social capital, producing a virtuous circle (Proost et al. 2009). The importance of social capital in agriculture emerges when social relations also have an economic value and can help the realization of economic activities. Social capital was a theme of sociological debate until the 1960s (Bourdieu 1980; Coleman 1992; Granovetter 1985), but only since the 1990s has it become the object of analysis by economists (Tsai and Ghoshal 1998; Landry et al. 2002; Subramaniam and Youndt 2005; Becker and Tomes 1986; Becker and Murphy 2000; Barrutia and Echebarria 2010; Evans and Syrett 2007) and political scientists (Putnam 1993, 1995; Fukuyama 2001; Piselli 2007; Bjørnskov and Mannermar Sønderskov 2010). The best known description defines social capital as: "The trust, norms which regulate it, networks of civic associations, elements that improve the efficiency of society by promoting initiatives taken by mutual agreement" (Putnam 1993, p. 196). In fact, social capital is an indicator of the quality and intensity of the participation by an individual and/or small community to various informal networks or formal associations. In terms of economic theory, the neoclassical approach situates social capital in microeconomic analysis and treats it as a new factor of the production function along with physical capital, natural and human ones (Becker and Tomes 1986; Becker and Murphy 2000). A macroeconomic approach instead stresses the involvement of social capital in the development rather than growth process and assigns, in this way, a central role to meta economic inputs in development processes (Knack and Keefer 1997). The Italian literature on social capital focused on some issues, such as social capital and analysis of local systems, social capital and innovation or social capital and network analysis (Cecchi et al. 2008). The socially oriented market niches are the ideal environment for both the sedimentation of social capital and for the birth of a new consumer. He may be a "critical" consumer—when he focuses on the negative aspects of the production models—a "responsible" consumer—when he is interested in knowing the social cost of the products—a "conscious and ethical" consumer—if he cares about the welfare of third parties. For this new type of consumer the quality of goods and services also includes the consideration of ethical choices and social responsibility of firms. The latter is therefore an indicator of social capital accumulated in a network. The literature on corporate social responsibility began in the 1960s (Frederick 1960), and evolved until the 1990s (Carroll 1999, Maon et al. 2009). In the 1990s, in fact, there was one of the most acknowledged descriptions of social responsibility: "the social responsibility of business encompasses the economic, legal, ethical and discretionary [later referred to as philantropic] expectations that society has of organizations at a given point in time" (Carroll 1979, p. 499). Over the past 30 years, social responsibility has had increasing consideration, probably due to the development of globalization. In fact, the control of the stakeholders on the activities of enterprises has greatly increased, and, at the same time, the need to ensure the health of ecosystems, social equity and good corporate governance has emerged. The initiatives taken on this issue are inspired by the concept of the triple bottom line, which includes social, economic and environmental sustainability (Kitzmueller and Shimshack 2012). More and more firms believe that reporting their initiatives in relation to this triple approach is significant for their reputation. They seek to reduce their ecological footprint and choose to achieve a reputation expendable for their sustainable organization. For these firms, social responsibility is preferable to the alternative of trying to elude the application of environmental policy. In this regard, also the European Commission Communication 681 (2011) stresses the importance of social responsibility for competitiveness in international markets: "… a strategic approach towards the issue of corporate social responsibility […] can bring benefits in terms of risk management, cost reduction, access to capital, customer relations, human resources management and innovation skills". Greater interest in this issue has developed recently in economic and agricultural areas. (Di Iacovo et al. 2005; Briamonte and Hinna 2008; Peri 2008; Inea 2007; Pulina 2011). Indeed, the environmental and social issues have led to major changes in the needs of citizen-consumers, in the policies, and in the strategies of farms. The rules and institutional constraints aimed at solving these problems, initially represented additional costs and constraints for farms and had a profound effect on their competitive capability. The European agricultural sector has replied to these circumstances with their own peculiarities and characteristics. They produced an agricultural multifunction model capable of responding to the different and changing needs of society. The farms are particularly interested in social responsibility, given their impact on the planet, biodiversity, climate and the local economy. Farms follow social responsibility on a purely voluntary basis, and not as a result of legal or formal obligations (Briamonte 2010; Carroll and Shabana 2010). The adoption of practices implies the interaction between firms and environment, and also means additional costs due to increased investments, the increased use of resources and the restriction of strategic choices. The compensations are in the improvement of the reputational capital, thanks to a wider stock of intangible resources, both internal, such as human and relational capital, or external, such as respect and the sense of belonging. However, with reference to the specific peculiarities of Italian agriculture it seems difficult to combine social responsibility with the world of food and, above all, with the small size of typical Italian farms. To this end, there have been two paths of action: in some cases the choice to adopt sustainable behaviour was the result of external pressure from the stronger players in the supply chain (Hartmann 2011). At other times, however, agricultural enterprises have decided their governance on the viewpoint, values and typical contents of social responsibility without being aware, regardless of network relationships. Therefore farms, as well as generating value through the production of goods and services required by society, aspire to realize a wide-ranging and stable well-being (shareholder value), implementing socially responsible behaviour. Over time, social responsibility has become a strategic factor in the process of innovation, a real competitive development strategy, based on the principle that farms should pursue not only economic but also social objectives. In this way the farm attains long-lasting advantages and repositions itself from its competitors. The gradual transition from traditional models of agriculture to those oriented to social responsibility took place in the farm through multifunctional agriculture, which nowadays plays an increasing social function linked to ethical paradigms and welfare. In fact, in addition to the usual goods of the primary sector, the farm is also able to produce positive externalities and public goods related to environment and food security. This has promoted a new entrepreneurial view and particular attention by the farm not only to the consumer but also to society. In practice, the emergence of a collective awareness on the necessity for responsible behaviour on the part of all stakeholders of the economic system has led to a new model of farms, more careful to incorporate the environmental and social concerns into production and commercial policies. The orientation towards social responsibility by farms designs competitive scenarios strategically characterized by strong ethical values. Indeed, the farm also includes ethical issues in its objective function, so it can respond to the expectations of its stakeholders. It achieves this purpose by adopting multifunctional and socially responsible competitive strategies. Therefore, the competitiveness is no longer based on product differentiation, but on the specific qualities and reputational capital arising from the practice of social responsibility. The sustainability and the protection of the consumer become the new competitive factors of this model of firm. In the agricultural sector, the multifunctional and multi-value farm is founded on entrepreneurial capital. The new models of entrepreneurship are the strategic input necessary to direct the processes of formation and accumulation of value and to manage the network and their governance structures. The ethical capital that characterizes these farms represents the key competitive strategy of the advanced countries to tackle the environmental and social dumping generated by globalization.

7 The "New" Public Agricultural Policies

In response to the various social concerns, agriculture is expanding its role from one of mere producer to a provider of goods necessary to the community. In this way the sector integrates an ethical value, though not always noticeable and not fully recognized. The theories mentioned above on Corporate Social Responsibility reflect on economic agents who, continuing to pursue profit, voluntarily submit themselves to additional constraints to meet the demands of the different stakeholders, and on economic relations based on social, environmental compliance. Many environmentalists believe economic theory is responsible for environmental problems, as it identifies development with the growth of availability of material goods but it seems to ignore the repercussions on quality of production processes and on consumption patterns. In fact the "traditional" economic theory was founded and developed in a world where the problems of pollution and environmental degradation were not as relevant as today. Therefore it ignores any economic problem arising from the environment. In addition, the economic literature cannot accurately quantify the effects produced by the accumulation in the long term of externalities not considered, in other words, the "intergenerational transmission of externalities".

In general, economic problems occur when social and private interests diverge. Pigou (1920) already showed how the existence of a gap between social costs and private ones depend on the allocation of resources as determined by the market. According to Georgescu-Roegen (1979), the market mechanism was never capable of solving the environmental problems, so in this field a failure of the market is produced. From this point of view, economic theory legitimizes public intervention to promote a new paradigm in agriculture, and especially in supporting innovation, with the need to remove the "market failures". The concept of sustainable development, which derives from these issues, is the basis of the choices of innovation and organization of farms. Over time the agricultural policy strategically enlarged its sphere of intervention from food security and poverty decline to the encouragement of research and development of small farmers, to the institution of rules and safety standards and to administration of public resources. Public intervention acquires a new justification in the necessity to respond to the constraints that create an environment which is not advantageous to innovation. In fact, the automatic market incentives are inadequate to stimulate firms to invest in new processes and new products to the required level. Most rural development models are based on the active policy of the State. The models were different: some were characterized by the prevalence of the environmental aspects on those properly productive. They related to multi-functionality and production diversification converged in a "New Theory of Rural Development" (NTSR) (Van der Ploeg et al. 2004) or they are referred to as a real "Model Territorial Production". In them the material resources (biodiversity, typicality) and intangible ones (tacit knowledge, traditions, and identity) are inputs of innovative processes of value creation. In all these models there was a separation between the entrepreneurial function—to contribute to the creation of national wealth—and the role of the state—protection of the interests of the weakest and the social and territorial cohesion. But the function of public administration is to transform (EC 2001; OECD 2008; Kaufmann et al. 2008). It is no longer the only source of decision-making power, but must direct and supervise complex relational networks of the territory where it works, involving the highest number of civil society stakeholders in decision-making and reallocating the power between different subjects. Following this approach, public administrations at all levels should be transparent, efficient and effective, requisites which requires involving and coordinating all stakeholders through the implementation of innovative governance apparatus. In fact, good governance has recently been introduced among the four basic factors that guarantee a better quality of life in rural areas, together with the environment, the economy, the services and the social and cultural capital (EENRD 2010). In practice, the public good nature of agricultural activity involves a different concept of Agriculture and Rural Development policies. They will consist of a special contract between farmers and society, according to which the production will receive public support only if it grants an efficient and sustainable use of natural resources. This implies that the link between producer and consumer will strengthen more and more over time. In its intense way, this relation leads to the formation of a more independent local economy that absorbs the consumer in production strategies and in the division of production risks. In other cases, these relationships induce the farms to the diversification of activities to satisfy a wider range of requests, according to a multi-functional concept of social farms. This diversification of agricultural production requires consumers who are aware and responsible. Through direct contact with the farm, they increase their knowledge of the production and so can benefit from both the agricultural production, in the strictest sense, and from positive externalities connected to public goods. In this way a virtuous circuit is activated, because it increases the motivation of consumers to pay as they accredit a premium price to goods and services acquired which, in turn, incorporate the value of intangible public goods that are part of the production process (Marotta and Nazzaro 2012). The crisis of world agricultural markets in 2008 produced a significant growth in agricultural prices, which has generated a dispute on agriculture and the policies dedicated to it, both in terms of objectives and of instruments. Opinions are completely divergent: on the one hand, some consider the increase in prices as an opportunity to clear all agricultural support policies, on the other hand, others argue that the crisis in question has highlighted the necessity of mechanisms to support and stabilize markets in a strategic sector such as agriculture, where an excessive price oscillation can lead to damaging consequences in terms of food security. The gradual reduction of the role of the state as a result of the sovereign debt crisis is causing a negative impact on economic and social life, and threatens the survival of instruments of public welfare and the social cohesion in the territories and between social classes. This problem can be mitigated in part thanks to new production models capable of producing, at the same time, economic and social value. Responsible agricultural practices which are related to the civilian economy, can render environmental and social sustainability a subsidiary aim to economic sustainability. They therefore represent a possible solution to re-establishing the paths of rural development. The gradual contraction of the financial resources available for agriculture forces them to pursue an efficient support for the benefit of the production of public goods. In fact, the need for a better environment for the community and, more specifically, the need for a more attentive use of the land, represent reputable reasons for agriculture support. The agricultural activity carried out by farms in the territory also pursues tasks related to producing public goods, i.e. goods characterized by two features: non-rivalry and non-excludability. In fact, agriculture is the unique economic activity that utilizes resources and produces a variety of goods and services which are, at the same time, assets owned by farmers and assets of a public nature for the community. The need to stimulate the initiatives of the CAP in promoting the production of public goods is an issue widely discussed at Community level with numerous contributions by the scientific community, environmental organizations and other stakeholders (ELO-BirdLife 2010). In the 1960s, the agricultural sector was already considered not only the expression of an entrepreneurial activity, but also a kind of public service. For its part, the European Union, with the introduction of set-aside and the "decoupling", introduced the principle that farmers cannot be compensated for what they produce, but for their role in the preservation of ecological systems. This consideration of the role of the agricultural sector highlights its natural implication; it cares about the environment and landscape, rather than the economic and productive aspects. From this point of view, the concepts of multi-functionality and multi-sector have a strategic importance and they are now the basis of many agricultural strategic choices. The relevance of this theme has been corroborated by a recent resolution of the European Parliament (2011), where it is stressed that in the reformed CAP, "public funds should be recognized as a legitimate form of payment for public goods, provided to the community, the cost of which are not offset by market prices". In Italy, however, there are recurring constraints on which public policy can only work in the long run: the limited resources, the small size of farms, which hinders the advance of a real internal research, the rising complexity of knowledge required for agricultural activities, the dominance of innovations imported from other sectors and other countries, since agriculture is a sector "dependent on suppliers" for the needs of innovation (Pavitt 1984). In this respect, it should be noted that according to data of the last General Agricultural Census (2010) there have been significant transformations: the average farm size (in hectares of UAA per farm) is larger, which demonstrates that Italy is approaching the European average. Small farms have decreased significantly, although those with less than 2 hectares still represent 51 % of Italian farms. However, the remarkable decline in small farms would suggest a defragmentation process of Italian agriculture and an orientation towards the European average. Human capital endowment will increasingly represent a major factor in the evolutionary processes of social change in the agricultural sector. Human capital plays a significant role in the new paradigm of agriculture, also recognized in the literature; in fact it affects the survival and growth of farms, their investment decisions (Huffman 1980) and their productivity. In recent years this concept has become increasingly important, thanks to the recognition of the leading role played by knowledge-intensive activity. It represents an important qualitative aspect of the job, plays a key role in determining rates of domestic investment and local entrepreneurship, and has the ability to generate or absorb innovations. Therefore, it has a chain effect on economic activity and employment and hence the growth of a country. Training, education and consulting services contribute to the enhancement of human capital in order to pursue the objective of competitiveness. Therefore, the ability to change and reorganise agriculture and farms is closely related to entrepreneurial dynamism, typical of the young generation.

8 Conclusions

Since 2008 the consequences of the global crisis in Italy have led to a serious contraction in output, employment and consumption, which have brought about structural changes to the economy of the country. In recent years, the contribution to economic growth (GDP) and employment by the agricultural sector were higher than those achieved by manufacturing. In addition, the added value contributed by the food industry is steadily higher than the industrialy average. These encouraging performances appear to be related to the important innovations that have recently characterized Italian agriculture. The key features are: the coordinated strategy of new kinds of organization of horizontal and vertical integration designed to re-balance the power of upstream and downstream stages; the improvement of productivity through structural investments; new rules and contractual arrangements for the placement of products in the early stages of the supply chain; the different product mix with a strong identity and quality improvement; the diversification of farm activities (agro-tourism, environmental services, energy production, etc.), which have served to achieve greater resilience to the shock. The diversification of revenues in this way allows farms to reduce the negative effects of the crisis, especially in the areas most affected by price volatility. A multifunctional farm can more easily offset the negative effects of a crisis through a reallocation of production factors from agricultural production to environmental and social tasks, in response to the new demand for goods and services expressed by society to the primary sector. Often farms with supplementary activities (agro-tourism, the processing of farm products, certified production) are less affected by the crisis than ones with undifferentiated products. They, therefore, endeavour to find new ways to build a system and find synergies to enhance the output of the respective value chains. They cooperate with each other to create value, but at the same time compete on returns of the value created. Despite the fact that the Istat national census in 2010 illustrated a predominantly family-based farming enterprise, in which the generational replacement continues with some difficulty, there are still many situations of vitality and innovation, with numerous farms directed by women or young people or oriented towards diversification, which are more attentive to the territory and the environment. The real portrait is therefore more composite and multifaceted, because in Italy agricultural activities (for internal consumption, for leisure or social activities) with small acreage and low income, but which are capable of generating economic, environmental and social impacts proliferate, though they are not always identified in the official statistics. The Italian agricultural sector is still underpowered, but it has a lot of possibilities and forms. It is characterized by new organizational models, the most widespread and important of which is the network. The networks between firms are interesting because they involve an almost natural series of mutual obligations for the companies themselves. The enforcement of these obligations lies—more than in the formal and legal instruments—in the close relationship that it creates between the success of the network and the success of individual companies. Thus, they have an interest in contributing to the success of the network and to refrain from opportunistic behaviour. Within this organizational model the transformations of food chains, changes in agricultural policy, the progressively closer link between agriculture and land use, and multi-functionality stand out. This progression makes the boundary between agriculture and other economic sectors less defined. In fact, traditional agriculture can no longer resist, but must focus on multifunctional choice. In addition to producing and selling commodities, the agricultural system must be able to interrelate with tourism, with customs, with renewable energy. It is important to evaluate which choices may be more successful in the future: a potential growth path for European agriculture could be the "economy of the experience", according to which consumers are willing to pay more for products and services that provide additional intangible 'experiences'. In this sense, the products of the agricultural sector include intangible assets, such as specific know-how, image, skill and custom. In this respect, the Italian agricultural sector is a European and global leader and the strong "Made in Italy" brand is one that helps the allocation of its products abroad. Nevertheless, innovative features such as multifunctionality and agricultural diversification, young farmers and the establishment of new employment in agriculture are very promising, but they must be delivered with new theoretical and practical models. Such a model should facilitate the unions between producers and ensure the product is not considered only as a commodity but as an asset that has added value and provides income. The new paradigm that is emerging in agriculture has different, economic, ethical and social foundations. While the competitive paradigm focuses on the concept of rivalry between firms (Porter 1990) the new paradigm is based on the strategic alliance, on the ability of firms to network and maintain stable relationships that create a relational advantage. Rural development will depend on how the various agricultural players will be able to interact. The new agricultural paradigm also implies a new agricultural governance based on the principles of dialogue, agreement, inclusion, participation, involvement, cooperation, networking, coordination, multi-sector and responsibility. In this way agriculture becomes a "system" which, despite partisan disagreements, is able to prevail and to reinforce its status in the supply chain. Therefore, needs public support and the public authorities have an important role because they can implement the right steps and measures to make the farms stronger and correct their competitive weakness compared to international competitors. Despite the fact that Italian agricultural research and experimentation is very dynamic, relationships and coordination between research, technology transfer and enterprises are not completely efficient and advantageous. To this end, universities can play a strategic role. They, in fact, operate in three areas simultaneously—research, teaching and technology and knowledge transfer—and can be a relevant factor in the processes of coordination throughout the entire supply chain of knowledge.

References

Aguglia L, Henke R, Salvioni C (2008) Agricoltura multifunzionale. Comportamenti e strategie imprenditoriali alla ricerca della diversificazione. Inea Studi & Ricerche, ESI, Napoli

Barrutia JM, Echebarria C (2010) Social capital, research and development, and innovation: an empirical analysis of Spanish and Italian regions. Eur Urban Reg Stud 17(4):371–385. doi:10.1177/0969776409350689

Battershill MRJ, Gilg AW (1998) Traditional low intensity farming: evidence of the role of vente directe in supporting such farms in Northwest France, and some implications for conservation policy. J Rural Stud 14(4):475–486. doi:10.1016/S0743-0167(98)00011-4

Becker GS, Tomes N (1986) Human capital and the rise and fall of families. J Labor Econ 4(3):1–39. doi:10.1086/298118

Becker GS, Murphy KM (eds) (2000) Social economics. Market behavior in a social environment. Harvard University Press, Cambridge

Bjørnskov C, Mannermar Sønderskov K (2010) Is social capital a good concept? In: 67th annual national meeting of the Midwest Political Science Association, Chicago. doi:10.1007/s11205-012-01991

Bonanno A (2012) Functional foods as differentiated products: the Italian yogurt market. Eur Rev Agric Econ 40:1–27. doi:10.1093/erae/jbr066

Bourdieu P (1980) Le capital social. Idées économiques et socials 3:63. doi:10.3917/idee.169.0063

Briamonte L, Hinna L (2008) La responsabilità sociale per le imprese del settore agricolo e agroalimentare. Esi, Napoli

Briamonte L (2010) La Responsabilità Sociale nel sistema agroalimentare: quali prospettive? Agriregionieuropa 20:88–99

Brunori G, Rossi A, Guidi F (2012) On the new social relations around and beyond food. Analysing consumers' role and action in Gruppi di Acquisto Solidale (Solidarity Purchasing Groups). Sociol Rural 21:1–9. doi:10.1111/j.1467-9523.2011.00552.x

Callon M (1999) The role of lay people in the production and dissemination of scientific knowledge. Sci Technol Soci 4:81–94. doi:10.1177/097172189900400106

Capitanio F, Coppola A, Pascucci S (2009) Indications for drivers of innovation in the food sector. Br Food J 8:820–838. doi:10.1108/00070700910980946

Capitanio F, Coppola A, Pascucci S (2010) Product and process innovation in the Italian food industry. Agribus Int J 26:503–518. doi:10.1002/agr.20239

Carroll AB (1979) A three-dimensional conceptual model of corporate performance. Acad Manag Rev 4(4):497–505. doi:10.5465/AMR.1979.4498296

Carroll A (1999) Corporate social responsibility—evolution of a definitional construct. Bus Soc 38(3):268–295. doi:10.1177/000765039903800303

Carroll AB, Shabana KM (2010) The business case for corporate social responsibility: a review of concepts, research and practice. Int J Manag Rev 12(1):85–105. doi:10.1111/j.1468-2370.2009.00275.x

Cecchi C, Grando S, Sabatini F (2008) Campagne in sviluppo: capitale sociale e comunità rurali in Europa. Rosenberg and Sellier, Torino

Cicatiello C, Silvio F (2008) La vendita diretta: produttori, consumatori e collettività. Agriregionieuropa 14:98–115

Coleman JS (1992) The problematics of social theory. Theory Soc 21(2):262–279. doi:10.1007/BF00997791

Di Iacovo F, Senni S, De Knegth J (2005) Farming for health in Italy. In: Hassing J, Elings M (eds) Farming for health across Europe. Frontis and Springer, UK

Dormon E, Leeuwis C, Fiadjoe FY, Sakyi-Dawson O, van Huis A (2007) Creating space for innovation: the case of cocoa production in the Suhum-Kraboa-Coalter district of Ghana. Int J Agric Sustain 5(2&3):232–246

European Commission (2001), Libro verde "Promuovere un quadro europeo per la responsabilità sociale delle imprese". http://ec.europa.eu/green-papers/

European Commission (2011), Commission staff working paper, impact assessment. Common agricultural policy towards 2020, assessment of alternative policy options. http://ec.europa.eu/agriculture/policy-perspectives/impact-assessment/cap-towards-2020/report/annex2_en.pdf

EENRD (2010) Working paper on capturing impacts of leader and of measures to improve quality of life in rural areas. Evaluation expert network for rural development Brussels: DG for agriculture and rural development. http://enrd.ec.europa.eu/evaluation/evaluation-of-rdp-2007-2013/evaluation-guidelines/en/evaluation-guidelines_en.cfm

ELO-BirdLife (2010), Proposals for the future CAP: a joint position from the European Landowners' Organisation and BirdLife International. http://www.birdlife.org/eu/EU_policy/Agriculture/eu_agriculture_ELO_joint_proposal.html

Evans M, Syrett S (2007) Generating social capital? The social economy and local economic development. Eur Urban Reg Stud 14(1):55–74. doi:10.1177/0969776407072664

Feder G (1980) Farm size, risk aversion and the adoption of new technology under uncertainty. Oxf Econ Pap New Ser 32(2):263–283

Frederick WC (1960) The growing concern over business responsibility. Calif Manag Rev 2:54–61. doi:10.2307/41165405

Fukuyama F (2001) Social capital, civil society, and development. Third World Q 22(1):7–20. doi:10.1080/713701144

Georgescu-Roegen N (1979) Energy analysis and economie valuation. South Econ J 45:1023–1058. doi:10.2307/1056953

Granovetter M (1985) Economic action and social structure. The problem of embeddedness. Am J Sociol 91:481–510. doi:10.1086/228311

Griliches Z (1957) Hybrid corn: an exploration in the economics of technological change. Econom 25(4):501–502. doi:10.2307/1905380

Hartmann M (2011) Corporate social responsibility in the food sector. Eur Rev Agric Eco 38(3):297–324. doi:10.1093/erae/jbr031

Hassanein N (2003) Practising food democracy: a pragmatic politics of transformation. J Rural Stud 19(1):77–86. doi:10.1016/S0743-0167(02)00041-4

Hinrichs C (2003) The practice and politics of food system localization. J Rural Stud 19:33–45. doi:10.1016/S0743-0167(02)00040-2

Huffman WE (1980) Farm and off-farm work decisions: the role of human capital. Rev Econ Stat 62(1):14–23. doi:10.2307/1924268

Inea-Oiga (2005) Insediamento e permanenza dei giovani in agricoltura. Gli interventi a favore dei giovani agricoltori. Rapporto2003/2004. Inea Roma. www.inea.it

Inea (2007) Promuovere la responsabilità sociale delle imprese agricole e agroalimentari. Linee guida Inea 2007. www.inea.it

Inea-Oiga (2009). Insediamento e permanenza dei giovani in agricoltura. Le misure per i giovani agricoltori nella Politica di sviluppo rurale 2007–2013. Inea Roma. www.inea.it

Janssen S, van Ittersum MK (2007) Assessing farm innovations and responses to policies: a review of bio-economic farm models. Agricu Syst 94(4):622–636. doi: 10.1016/j.agsy.2007.03.001

Kaufmann D, Kraay A, Mastruzzi M (2008) Governance matters VII: aggregate and individual governance indicators 1996–2007. World Bank Policy Research WP 4654. doi:10.2139/ssrn.1148386

Kirwan J (2006) The interpersonal world of direct marketing: examining conventions of quality at UK farmers' markets. J Rural Stud 22(3):301–312. doi:10.1016/j.jrurstud.2005.09.001

Kitzmueller M, Shimshack J (2012) Economic perspectives on corporate social responsibility. J Econ Lit 50(1):51–84. doi:10.1257/jel.50.1.51

Knack S, Keefer P (1997) Does social capital have an economic payoff? A cross-country investigation. Q J Econ 112(4):1251–1288. doi:10.1162/003355300555475

Knickel K, Brunori G, Rand S, Proost J (2009) Towards a better conceptual framework for innovation processes in agriculture and rural development: from linear models to systemic approaches. J Agric Educ Ext 15(2):131–146. doi:10.1080/13892240902909064

Landry R, Amara N, Lamari M (2002) Does social capital determine innovation? To what extent? Technol Forecast Social Change 69:681–701. doi:10.1016/S0040-1625(01)00170-6

Maloni MJ, Brown ME (2006) Corporate social responsibility in the supply chain: an application in the food industry. J Bus Ethics 68(1):35–52. doi:10.1007/s10551-006-9038-0

Maon F, Lindgreen A, Swaen V (2009) Designing and implementing corporate social responsibility: an integrative framework grounded in theory and practice. J Bus Ethics 87:71–89. doi:10.1007/s10551-008-9804-2

Marotta G, Nazzaro C (2012) Responsabilità sociale e creazione di valore nell'impresa agroalimentare: nuove frontiere di ricerca. Economia Agro-Alimentare 1, Franco Angeli, Milano

OECD (2008). Environmental performance of agriculture in OECD countries since 1990, OECD Paris. http://www.oecd.org/agr/policy

Pascucci S (2010) Governance structure, perception and innovation in credence food transactions: the role of food community networks. Int J Food Syst Dyn 1(3):224–236

Pavitt K (1984) Sectoral patterns of technical change: towards a taxonomy and a theory. Res Policy 6:343–351. doi:10.1016/0048-7333(84)90018-0

Peri I (2008) Responsabilità sociale d'impresa, agricoltura e ambiente: implicazioni e applicazioni. In: Briamonte L, Hinna L (eds) La responsabilità sociale per le imprese del settore agricolo e agroalimentare. Esi, Napoli

Pigou AM (1920) Economics of welfare. Macmillan and Co, London

Piselli F (2007) Communities, places, and social networks. Am Behav Sci 50(7):867–878. doi:10.1177/0002764206298312

Porter ME (1990) The competitive advantage of nations. MacMillan, New York

Powell WW, DiMaggio PJ (1991) The new institutionalism in organizational analysis. The University of Chicago Press Book, Chicago

Proost J, Brunori G, Fischler M, Rossi A, Sumane S (2009) Knowledge and social capital. In: Innovation processes in agriculture and rural development. Final report of "IN-SIGHT: Strengthening Innovation Processes for Growth and Development". FP6

Pulina P (2011) Impresa agricola familiare, capitale umano e mercato del lavoro. Franco Angeli, Milano

Putnam RD (1993) Making Democracy Work. Civic traditions in modern Italy. Princeton University Press, Princeton NJ

Putnam RD (1995) Bowling alone: America's declining social capital. J Democr 6(1):65–78. doi:10.1353/jod.1995.0002

Röling NG (1992) The emergence of knowledge systems thinking: A changing perception of relationships among innovation, knowledge processes and configuration. Int J Knowl Transf Util 5:42–64. doi:10.1007/BF02692791

Röling NG (2009) Pathways for impact: scientists' different perspectives on agricultural innovation. Int J Agric Sustain 7:83–94. doi:10.3763/ijas.2009.0043

Ross RB, Westgren RE (2009) An agent-based model of entrepreneurial behavior in agri-food markets. Can J Agric Econ 57:459–480. doi:10.1111/j.1744-7976.2009.01165.x

Rossi A, Brunori G (2010) Drivers of transformation in the agro-food system. GAS as co-production of alternative food networks. In: Darnhofer I, Grötzer M (eds) Building sustainable rural futures. The added value of systems approaches in times of change and uncertainty. 9° European IFSA Symposium, Universität für Bodenkultur, Vienna

Ruttan VW (1996) What happened to technology adoption-diffusion research? Sociol Rural 36:51–73. doi:10.1111/j.1467-9523.1996.tb00004.x

Sckokai P, Moro D (2009) Modelling the impact of the CAP single farm payment on farm investment and output. Eur Rev Agric Econ 36 (3):395–423. doi:10.1093/erae/jbp026

Sini MP (2010) The importance of obtaining a more balanced relationship between the long and short food chain in the worldwide market for farm and food produce. 116th Seminar Eaae, Parma, Italy

Smith A (1776) An inquiry into the nature and causes of the weakth of nations. Straham and Caddell, London

Spielman DJ, Ekboir J, Davis K, Ochieng CMO (2008) An innovation systems perspective on strengthening agricultural education and training in sub-Saharan Africa. Agric Syst 98:1–9. doi:10.1016/j.agsy.2008.03.004

Subramaniam M, Youndt MA (2005) The influence of intellectual capital on the types of innovative capabilities. Acad Manag J 3:450–463. doi:10.5465/AMJ.2005.17407911

Sunding D, Zilberman D (2001) The aricultural innovation process: research and technology adoption in a changing agricultural sector. In: Gardner BL, Rausser GC (eds) Handbook of agricultural economics. Elsevier, Amsterdam

Teece DJ (1996) Firm organization, industrial structure, and technological innovation. J Econ Behav Organ 31:193–224. doi:10.1016/S0167-2681(96)00895-5

Tregear A (2011) Progressing knowledge in alternative and local food networks: critical reflections and a research agenda. J Rural Stud 27:419–430. doi:10.1016/j.jrurstud.2011.06.003

Tsai W, Ghoshal S (1998) Social capital and value creation: the role of intra-firm networks. Acad Manag J 41:464–476

van der Ploeg JD, Bouma J, Rip A, Rijkenberg FHJ, Ventura F, Wiskerke JSC (2004) On regimes, novelties and co-production. In: Wiskerke JSC, van der Ploeg JD (eds) Seeds of transition, essays on novelty production, niches and regimes in agriculture. Royal Van Gorcum, Assen, the Netherlands

van der Ploeg JD (2006) Structure and agency. In: Clark AD (ed) The elgar companion to development studies. Edward Elgar Cheltenham, UK

van der Ploeg JD (2008) The new peasantries: struggles for autonomy and sustainability in an era of empire and globalization. Earthscan, London

Vetter H, Karantininis K (2002) Moral hazard, vertical integration, and public monitoring in credence goods. Eur Rev Agric Econ 29 (2):271–279. doi:10.1093/eurrag/29.2.271

Welsh J, MacRae R (1998) Food citizenship and community food security. Can J Dev Stud 19:237–255. doi:10.1080/02255189.1998.9669786

Wilson GA (2008) From weak to strong multifunctionality: conceptualising farm-level multifunctional transition pathways. J Rural Stud 24:367–383. doi:10.1016/j.jrurstud.2007.12.010

World Bank (2006) Enhancing agricultural innovation: how to go beyond the strengthening of research systems, Washington. www.worldbank.org/rural

Chapter 3
Tourism: A Service Sector

Abstract The tourism sector is a large and fast growing industry able to generate several potential benefits for the whole economy both at local and national level. In recent years new forms of tourism (labelled as alternative) have emerged. Among these, rural tourism plays an important role due to its influence on the development of local communities. Tourism is an economic sector in which a high degree of entrepreneurial involvement is required. Moreover in order to compete worldwide, both entrepreneurship and networking emerge as key features of successful tourism provision. The aim of this discussion is to highlight the role of these features with particular reference to their influence on achieving a robust and resilient performance of the tourism sector that is required in the present economic crisis.

Keywords Tourism sector · Rural tourism · Entrepreneurship · Networks

1 Introduction

The service sector has become the most important economic sector, both in terms of GDP share and employment. Nowadays approximately 1/3 of global GDP is derived from services. In this sector labor productivity does not grow as fast as in agriculture and industry thus increasing the share of services in GDP. Moreover, services need a lower level of mechanization and a higher level of human capital compared to agriculture and industry where technological progress increases labour productivity and reduces employment.

Most services play a role in the production and marketing of goods and agricultural products by supporting the entire production process and providing value-added inputs for competitive industrial development.

This chapter is written by Annamaria Stramaglia.

A. R. Gurrieri et al., *Entrepreneurship Networks in Italy*,
SpringerBriefs in Business, DOI: 10.1007/978-3-319-03428-7_3,
© The Author(s) 2014

Tourism is one of the most important areas of the service sector, and, is frequently considered as a viable means of raising the economic activity of regions due to low entry barriers.

Moreover, it has been observed that the development of a tourism industry promotes the image of a destination enabling the region to achieve other objectives, such as an improvement in employment and a reduction in emigration.

The tourism sector possesses the inherent ability to create many potential benefits for the whole economy both at local and national levels.

Firstly, it is without question one of the foremost export sectors, having a higher capacity to attract investments and foreign capital. In addition, it serves to diversify the local economy. Improvement in this sector may provide a stimulus to the development of other economic sectors, namely agriculture, handcrafts, the agro-food industry and manufacturing.

Through tourism local entrepreneurial and management groups may also emerge, not only in the hotel industry but in other tourism related areas such as leisure activities, restaurants, nightclubs, water sport activities, shops, entertainment and ground transport. Tourism can generate jobs directly through its various activities, and indirectly through the supply of goods and services needed by tourism related businesses.

Moreover, it is important to stimulate development in local communities from a social point of view through their participation in tourism related events, but also in the reduction of migration, especially from rural areas.

This sector is also sensitive to extraordinary events both at national and international level. There are many positive examples, such as the Olympic Games or the World Cup and negative ones, such as political instability.

Tourism can be divided into mass and alternative tourism. The standardized mass tourism of the 1960s and 1970s was characterized by large numbers of people seeking holidays in popular destinations. It is being superseded by a new form of tourism driven by changes in consumer tastes, emphasising the tourist experience.

Within this new form of tourism, often defined as 'alternative', there are several specific forms such as cultural, educational, scientific, adventure and agro-tourism, with rural, ranch and farm subsets. It is one of the economic sectors which requires a high degree of involvement by entrepreneurial activity. More specifically, diversification of the services is necessary to cope with increasing demand for new types of tourism needs.

Additionally, entrepreneurship has a central role in managing all the other variables of the tourism system and is, therefore, a core aspect for its growth.

Global competition and the rapid progress in technology, communication and transport have created many new opportunities for tourism enterprises and at the same time increased the level of investment needed to benefit these opportunities.

It is clear that entrepreneurship and network, in their different forms, are two extremely important features of successful tourism provision. The different organizational forms, above all networks, emerge as the best way for tourism enterprises to compete worldwide.

We will focus our attention on rural tourism due to its important role as a potential economic development tool, particularly for rural communities. In many developing rural regions this form is increasingly being used and supported as a development strategy to improve social and economic wealth.

Tourism in rural areas provides alternative income and the preservation of local, natural values and culture for the local inhabitants. However, its potential influence on growth also refers to its ability to stimulate many related economic activities.

2 The Relevance of the Tourism Sector

The UNWTO (2012, p. 3) highlighted the importance of the tourism sector within the global economy summarizing the information from all countries with data available for 2008. It argues that "tourism's contribution to worldwide gross domestic product (GDP) is estimated at some 5 %. Tourism's contribution to employment tends to be slightly higher and is estimated in the order of 6–7 % of the overall number of jobs worldwide (direct and indirect). For advanced, diversified economies, the contribution of tourism to GDP ranges from approximately 2 % for countries where tourism is a comparatively small sector, to over 10 % for countries where tourism is an important pillar of the economy. For small islands and developing countries, the weight of tourism can be even larger, accounting for up to 25 % in some destinations".

In 2011, Europe accounted for over half of all international tourist arrivals worldwide, and it was the fastest-growing region on this front.

Within the European policy, the relevance of the tourist sector is clear from the very first paragraph of the European Commission Communication no. 352 (2010, p. 3): "Tourism is a major economic activity with a broadly positive impact on economic growth and employment in Europe".

It is an activity capable of generating growth and employment, while contributing to development and economic and social integration, with particular reference to rural and mountain areas, coastal regions and islands, outlying and outermost regions or those involved in convergence processes.

Furthermore, in the same Communication, the European Commission underlines the transverse nature of the tourism policy and the direct or indirect impact of many other European policies on the sector. Among others, EU rural development policy is of considerable importance to the tourism sector.

> Through the European Agricultural Fund for Rural Development (EAFRD), the Commission can support, among other things, the establishment of businesses active within rural tourism, the development and promotion of agri-tourism and capitalisation on the cultural and natural heritage of rural regions, including mountain areas (European Commission 2010, p. 14).

With reference to the developing areas, the World Tourism Organization, in its last Annual Report identified five sectors of particular importance: agri-food, information and communication technologies, textiles and apparel, tourism, and transport and logistics. Furthermore, as to the tourist sector, it argues that:

> tourism is a key services sector and an important source of revenue for developing and least developed countries, and yet it is still a largely untapped resource (WTO 2013, p. 99).

For developing areas, the tourism sector represents one of the main sources of foreign exchange income and the number one export category.

It is also very important for creating much needed employment and opportunities for development.

2.1 Evidence on the Performance of the Sector

Tourism is without doubt the world's largest and fastest growing industry. This is related to the emergence of modern welfare society, in which time consumption has gradually become an important activity for many people.

Over time the world tourism industry has had to face major challenges, which have also represented greater opportunities for expansion.

However, it has highlighted that the top 15 destinations in the world have, over the years, absorbed a lower percentage of all international tourism arrivals. In 1950 it was 98 %, in 1970, 75 %, falling to 57 % in 2007. This reflects the challenges for the sector, in particular to the emergence of new destinations, many of them in developing countries.

Furthermore, the sector has to adapt to social and economic development which has strongly influenced the evolution of tourist demand.

Also the economic context influences the performance of the tourist sector both at national and international levels. The European Commission (2010) pointed out that, as a consequence of the financial crisis, there was a fundamental change in the tourism sector. In particular Europeans began choosing different forms of travel. They have reduced the length of stay or their spending and are also more likely to choose closer destinations. This observation confirmed that, in spite of a slower pace than in the past, the tourism sector is still showing a healthy growth all over the world.

The distributional effects of the recent economic crisis clearly influence the results of the different tourist destinations and accommodations.

In Italy the performance of the sector in the last few years confirms the effect of the economic crisis, resulting in a profound structural change in accommodation supply. The most obvious effect is a significant reduction in internal demand, particularly for long stays. Recently this decline has become more and more important due to the increasing effects of the crisis on household income.

Several aspects of the sector structure have changed. The first, and also the most important, is the emergence and the development of new forms of accommodation. Furthermore, the substitution of the lower level of hotels by B&B, both in the urban and rural contexts, has also been important. Finally, the higher level hotels have developed, in order to meet the demands of wealthier tourism. This type of overall regeneration of the accommodation offer has enabled the tourist system to face up to international competition and maintain its market share in the extremely dynamic context of international tourism.

Despite falling domestic demand there is a positive trend in international tourism, driven, in particular, by the demand from recently developed extra-European countries. However, Italy seems to benefit from this new demand to a lesser extent compared to other European countries and to its direct competitors.

Based on Eurostat (2013), with reference to the outbound holiday trips made by EU residents in 2011, Italy was the second foreign destination. The top destination was Spain. The UNWTO (2012) performed the same analysis for all international tourist arrivals. Italy was confirmed in 5th position within the top destinations in both indicators used: arrivals (46 million) and receipts (US$ 43 billion). The top four destinations for international arrivals were, France, the United States, China and Spain.

These data highlight the difficulty of Italy in adopting a model of export-led growth in tourism. This is due to the significant weight that EU demand has on the overall budget. On the other hand, it shows the presence of sectors (particularly those related to new forms of tourism) capable of capturing international market shares sufficient to offset the negative effects of the decline in EU demand and thus making it less sensitive to the crisis.

3   New Emerging Forms of Tourism

Globalization has brought about significant changes to the tourism industry. These changes are influenced by economic factors but they also depend on the emergence of new cultures and a new generation of tourist.

Over the centuries tourism has also undergone profound changes due to the difficulty/ease of travel. It is a subset of travel, referring to the activity of visitors, a destination and travelling for recreational purposes.

There was a basic transformation in the tourism sector from Fordian Tourism to post-Fordian tourism, or rather, from mass tourism to a new tourism (cfr. among others: Poon 1993). The former represented the norm for more than five decades, the latter gave rise to the transformation of tourism demand that was initially based on the development of modern information and communication technology. In short, there has been a shift in emphasis from passive fun to active learning.

Alternative tourism includes various forms of tourism which are:

- Thematic tourism (Cultural and Religious tourism);
- Adventure tourism;
- Rural tourism.

As pointed out by Fayos-Solá (1996) the main characteristics of this new form are the segmentation of demand, the need for flexibility of the supply and distribution, the diagonal integration and subsequent system economies instead of economies of scale. Diagonal integration is a process in which the tourism enterprise can develop and compete in more than one activity, seeking synergies between different products and integrated services. Compared with vertical and horizontal integration, it is based on system economies rather than economies of scale. According to Poon (1993) the tourism sector has undergone profound changes that may be classified as follows: (1) new consumers, (2) new technologies, (3) new forms of production, (4) new management styles, and (5) new prevailing circumstances.

These changes are based on the globalization and the development of information and communication technology.

The new tourists are characterized by a higher level of experience and independence, they are also more interested in product quality and in environmental protection.

On the supply side the tourism enterprises have an increasing need for in-depth knowledge of the market in order to identify the emerging traits and needs. They also have to experiment new tourism products and new ways of communication and distribution. Furthermore, tourist enterprises have to became more flexible in several areas: organization, production, distribution, reservation and payment systems, for which new information and communication technologies are fundamental.

Within the new tourism forms Sharpley (2002) argues that rural tourism represents an important opportunity to diversify the product portfolio of declining mass tourism destinations. Moreover, it can contribute to enhancing the positioning of countries generally associated only with coastal tourism.

3.1 Rural Tourism in Literature

Although the European Commission has proposed one of the first definitions of "rural tourism" (European Commission 1986), this is vague and quite simplistic. In fact, any tourist activity that takes place in rural areas is considered rural tourism.

More recently it has been defined by the EU as a broader concept than accommodation. It also covers services such as events, festivities, outdoor recreation, production and sale of handicrafts and agricultural products. The intention of the European Commission is not to provide a unique definition or a normative framework for all the member states. Its main purpose is the protection of the environment and local traditions, and its only role is the planning of financial projects.

In the nineties, when rural tourism was beginning to assert itself, scholars had attempted to construct a theoretical framework in order to define and study its development. Among many others, these included the studies by Dernoi (1991), Frederick (1992), Oppermann (1996) and the many works of Bernard Lane.

The latter is considered ground-breaking in his attempt to define and clarify rural tourism. According to Lane, rural tourism exists as a concept, which is a form of tourism practiced in rural areas. This tourism form is rural in scale, character and function. Furthermore, it reflects the different and complex model of the environment, economy, history and rural location (Lane 1994).

Dernoi (1991) tried to distinguish between wild tourism and rural tourism.

According to the author, the former is the tourism which takes place in non-urban areas such as outdoor recreation in wild areas, national parks, national forests and generally uninhibited areas. The latter includes farm tourism, non-farm tourism in rural areas and communities.

Frederick (1992) underlines the typical characteristics of rural tourism that represent the difference from the conventional type. These are:

- the locations are sparsely populated;
- there is the predominance of the natural environment;
- it is experience oriented;
- it pays attention to the preservation of culture, heritage and traditions;
- it has a lot of potential benefits for rural areas.

Scholars have never agreed on the question that rural tourism represents a subset of tourism for which there is no unique definition (Busby and Rendle 2000; Roberts and Hall 2001). However, the most common approach is to define this subset as one with farms that offer lodging to their guests.

In particular, Roberts and Hall (2001) underlined the different terms that identify a unique phenomenon, so rural tourism, agro-tourism, farm tourism and even eco-tourism are often alternatives used to label tourism activities which take place in rural areas.

From a general point of view rural tourism is a multi-faceted activity. In fact it encompasses all activities which are undertaken in rural areas such as eco-tourism, adventure tourism, etc. (McGehee and Kim 2004).

Busby and Rendle (2000) collected 13 definitions chronologically in order to illustrate a gradual change of what is meant as rural tourism. They argue that the link between tourism and agriculture becomes progressively weaker, passing from tourism on farms to tourism in tourist companies. Thus the term rural tourism is preferred. It includes farm-based holidays but it has a much wider scope. For example, Lane (1994) also includes special interests in walking, climbing, and riding holidays, sport, adventure and health tourism, hunting and angling, educational travel, arts and heritage.

Cawley and Gillmor (2008) argue that tourism can be defined as rural if it is possible to identify strong links with the economic and productive activities that take place in the area.

Another important aspect is the presence of three main features: integration, sustainability and endogeneity.

As stressed by Farmaki (2012), the answer to the initial questions, and then the rural tourism definition, may regards three different aspects: (1) geographic and demographic; (2) product; (3) a tourist experience.

As to the product aspect, the key elements are the services offered, such as rural adventure tours, rural resorts, events and agricultural education, and so on. Finally, the definition of rural tourism is based on the tourist's experience of the products and activities of the area. Thus, the core elements to define rural tourism are:

- the difference from mass tourism,
- stress on individual/small group approach,
- sustainability,
- environment and nature protection,
- enhancement of traditions,
- enjoyment, but also education.

In particular, sustainability refers to the objective of minimizing tourism influence on the local environment, therefore sustainable tourism is a continuation of the concept of sustainable development.

According to Roberts and Hall (2004), the key characteristic that distinguishes rural tourism is the approach that focuses on the importance of supply management and marketing activities.

3.2 Rural Tourism and the Presence of Agri-tourism

Although in many contexts rural tourism and agri-tourism are treated as synonymous, it is possible to distinguish the two concepts. Rural tourism is a wider concept while agri-tourism may be considered as one of the forms of rural tourism, in other words it is just one variety of the whole spectrum.

Generally, however, it is difficult to specify a single definition either for agri-tourism or rural tourism.

Phillip et al. (2010) provide a comprehensive examination of the literature on the definition of agri-tourism and summarize the existing definitions and related labels.

Agri-tourism is a style of vacation in which hospitality is offered on farms. This may include the opportunity to assist with farming tasks during the visit. It is a tourism form which capitalizes on rural culture as a tourist attraction.

On the other hand, rural tourism includes both agri-tourism and other types of tourism experience. This is because, the simple EU definition of rural tourism is of a recreational experience not necessarily agricultural in nature. It involves visits to rural settings or rural environments for the purpose of participating in or experiencing activities, events or attractions.

Agritourism is a specific form of rural tourism with the following characteristics: (1) a direct relationship to agricultural activities or buildings with agricultural functions, (2) it is provided by entrepreneurs whose main activity is farming, (3) it represents an extra activity to acquire additional income.

These characteristics are included in the Italian definition of agritourism. Rural tourism is not just farm-based tourism, it may be offered by an entrepreneur who is not farmer, and it is more related to cultural or naturalistic experiences.

However, it is important to note that in most of the analyses these two concepts are considered synonymous.

As pointed out by Roberts and Hall (2001), rural tourism is not a new phenomenon. In fact rural areas have long provided the setting for recreation and tourism activities. However, throughout the last few decades a traditional niche market, such as rural tourism has become a very important and fast-growing industry for many areas of the world (cfr. among others: Park and Yoon 2009 for the Korean case; Chuang 2013 for the Taiwan case; Dimitrovski et al. 2012 for the Serbian case; Sharpley 2002 for the Cyprus case).

This vast expansion may be caused by the benefits that this kind of tourism offers to the host community, to the land and also to the tourist (San Martin and Herrero 2012). Furthermore, rural tourism is an efficient way to increase the income of rural inhabitants, and to reach other goals such as employment growth, repopulation, social improvement, and revitalization of local crafts.

The increased interest in this form of tourism with reference to developing countries is a result of the realization that a systematic effort is necessary to create better living conditions in the rural areas where the vast majority of populations of developing countries reside.

A rural tourism activity may be easily built on the existing resources on the farm and does not require any huge changes to the agricultural landscape.

The importance of rural tourism lies in its characteristics. It focuses on four dimensions: (1) social, (2) economical, (3) environmental, and (4) cultural. All of these also refer to the concept of sustainable development and this helps to develop the socio-economic situation both at individual and at regional level.

In European areas there have been many interventions intended to strengthen tourism and its competitiveness. Over the last few decades the European Commission has invested a lot of effort in implementing actions referring to the tourism sector (cfr. European Commission 2006, 2007, 2010).

One of these is the INTERREG IV project. It combines tourist entrepreneurship and agriculture in the EU new member states which try to stimulate collaborative experiences in the sector, also by way of educational activities, for example in rural design, traditional cuisine and marketing.

The preservation of rural areas and the development of economic activity aimed at reaching this goal represent an important aspect of European policy. Every policy action that may have a positive impact on the development of rural areas is relevant.

Also in non-EU countries it is possible to find several cases of public intervention. For example, with reference to Korea, Park and Yoon (2009) underline that the local government has played an important role through the promotion of two projects focused on the development of this niche sector. The main reason for these policies is to encourage 'bottom-up' development that moved just from the development of local cultural resources.

4 Entrepreneurship in Tourism

The fundamental role of small and medium-sized enterprises for job creation and economic growth has often been highlighted. At the same time it is known that the service sector employs the majority of employees in developed countries and, as such, both the tourism sector and entrepreneurship within it receive an increasing level of attention. From this point of view, tourism businesses represent essential actors, firstly for creating jobs, and generally for increasing the economy.

From the point of view of entrepreneurship, the tourism sector provides a particular context that differs from other economic ones. These are the identification of entrepreneurial opportunities and their conversion into a consumable tourism service.

In order to reach the growth goal of the whole economy, it is necessary to understand and highlight the importance of entrepreneurship and human resource management within the tourism sector.

Komppula (2014, p. 364) argues that: "… a major part of small tourism enterprises do not pursue growth: the low entry barriers in the tourism sector encourage the proliferation of micro and small firms, which may fail to appreciate the importance of all kinds of development of the firm, and many fail to recognize their dependence on the competitiveness of the destination as a whole".

The key to understanding entrepreneurship is to understand the entrepreneur. An entrepreneur is often defined as one who starts his own, new business. Despite this definition, several researchers (cfr. among others: Morrison et al. 2001; Morrison 2006; Wagener et al. 2010) agree that not every new small business is entrepreneurial or represents entrepreneurship. This is because, to be entrepreneurial an enterprise has to have special characteristics over and above being new.

From this point of view, entrepreneurs are a minority among new business owners. They create something new or different.

According to Schumpeter's approach, developed in the 1930s, the focus is on the entrepreneur as an innovator who is able to generate discontinuity through shaking up the established ways of doing things.

Moreover, Shane and Venkataraman (2000) highlighted the importance of discovery, evaluation, and exploitation of opportunities within the entrepreneurial process.

Bruyat and Julien (2000) state that entrepreneurs can be considered those who have a central role in creating new value. This new value may be either in the form of a new venture, or by bringing about significant changes in the ways of doing things.

This approach has been criticized, but underlines the individual's role and the difference between those who have entrepreneurial behaviours and those who simply run businesses. This difference is extremely important in the tourist sector in determining the economic and social impact.

The behaviour of entrepreneurs is important for different reasons:

- in determining the level of tourist satisfaction with a destination (Lerner and Haber 2001);
- in increasing the attractiveness of destinations.

On this second point researchers have demonstrated that even one able and creative entrepreneurial individual can stimulate the development of a destination (Johns and Mattsson 2005).

An entrepreneur's behavior arises from the presence of an entrepreneurial attitude towards entrepreneurship (that is towards innovation) and culture (Morrison 2006). The individual characteristics of an entrepreneur or their attitude towards entrepreneurship and an entrepreneurial culture are the main dimensions of their activity or behavior among tourism-related SMEs.

Tourism is a particular sector that provides a rich domain for creative entrepreneurs to create new cutting edge products and services. These are the expression of the entrepreneurs' view of what is needed in the contemporary tourism industry.

More recently, following the hypothesis that there is a difference between entrepreneurs and non-entrepreneurial owners in the hospitality industry, Wagener et al. (2010) empirically find that these are significant. More specifically, they reveal that there are several individual characteristics which discriminate between entrepreneurs and small business owners.

Among others, entrepreneurial owners have a psychological profile that is consistent with innovation, growth and profit, combined with the capacity to motivate other people. They possess higher levels of independence (i.e. relying on their own judgments), tolerance of ambiguity, risk-taking, innovativeness, and leadership qualities.

In addition, the authors found that entrepreneurs had more employees and higher profit while at the same time thinking about expanding their current venture.

Also Tajeddini (2010) shows that the magnitude of entrepreneurial orientation in the tourist sector has a significant and positive impact upon the achievement of profit and sales goals.

These results are important in order to identify the type of business ownership and its potential contribution to growth in the rural areas.

At an individual level, entrepreneurial behavior represents an expression of a developmental change by an individual. Thus, it is a key component in the sector development process.

The tourist sector is characterized by a high number of SMEs that are generally not able to appropriately adopt an entrepreneurial strategy.

In most cases tourist SMEs rely on social and family networking in order to raise capital, recruit employees and get business support (Johns and Mattsson 2005).

In this sector it is very important to increase the entrepreneurial orientation of the business owners as firms that are more adaptable and flexible, and which are characterized by higher levels of innovation and entrepreneurship, are successful also in turbulent conditions.

In the rural tourism sector the role of entrepreneurship is even more important due to the potential contribution of the sector to the growth of the rural areas.

This is because entrepreneurship can represent a tool, or at least a useful way of assessing tools, to further the development of tourist destinations.

In this context entrepreneurship is the ability to find a unique blend of resources outside of agriculture.

For example, this can be achieved by widening the base of a farm business to include tourism service related to agriculture. Thus, a rural entrepreneur is someone who enjoys staying in the rural area and contributes to local development. This is important because the economic goals of a rural tourism entrepreneur and the social goals of rural development are more strongly interlinked than in other non-rural contexts. For this reason entrepreneurship in rural areas is usually community based, has strong extended family ties and a relatively large impact on the community.

Peters et al. (2009) argue that a lot of tourist entrepreneurs, in particular rural ones, may be included in the definition of a lifestyle entrepreneur.

The lifestyle entrepreneur is the opposite of the growth-oriented and innovative entrepreneur (cfr. among others: Williams et al. 1989; Getz and Petersen 2005; Lashley and Rowson 2010).

According to the previous research several factors are associated with the phenomenon of the lifestyle entrepreneurs. The decision to start a tourist enterprise follows the desire to retain some control over working lives and on private leisure time. Moreover these entrepreneurs enjoy staying and working in an agreeable natural environment with together their family.

The tourism sector, and even more rural tourism, is characterized by the presence of both the classical Schumpeterian innovator entrepreneurs as well as lifestyle entrepreneurs who operate according to personal lifestyle.

Lifestyle entrepreneurs are often clearly motivated by non-economic reasons, even more so if they are in rural areas (Morrison 2006). They set up a business first of all to undertake an activity they enjoy themselves, then to achieve an activity that provides adequate income.

From an economic point of view this kind of entrepreneur accepts suboptimal levels of production, because an optimal level would have a negative effect on their quality of life.

Moreover, non-growth oriented entrepreneurs in the tourism sector (or non-economically motivated ones) represent a constrain for the development of tourism areas (Ateljevic and Doorne 2000).

Among others, the characteristics of lifestyle entrepreneurs that may be non-positively linked to the areas of growth are the limited use of resources and capital investment, limited marketing activities and use of information and communication technologies, and low involvement within industry structures.

Getz and Carlson (2000) have clustered two types of entrepreneurs in the Australian rural tourism and hospitality sectors: (1) family-first and (2) business first. The former cluster of entrepreneurs is motivated by emotional factors rather than economic ones. The prevalence of emotional motivation is associated with families and can be termed a lifestyle factor.

The presence of lifestyle entrepreneurs within tourism and the rural tourism sector highlights the importance of the policy makers in supporting the development of the tourist areas, the rural ones even more so.

More specifically, to our thinking, the policy maker should stimulate the development of entrepreneurial activities due to their influence on the recovery of rural tourism potential and regional traditions. Moreover, entrepreneurship may significantly contribute to maintaining local employment growth and increasing the living standards of the inhabitants.

Additionally, as the lifestyle entrepreneurs are not growth-oriented, because internal growth decreases the life quality, policy makers may stimulate processes of external growth rather than encourage internal growth. External growth through cooperation, clustering and/or strategic alliances should motivate entrepreneurs to reach an optimal level of production that represents a goal for the development of the area.

5 Networking in Tourism

In a globalized market context such as tourism, networks are an important form of inter-organization that allows integration of efficiency and stability with flexibility, innovation and creativity of small systems.

The collaborative relationships between firms ensure that the objective of growth and survival is achieved through the relational assets created through knowledge diffusion and sharing of the vision, mission and values of the organization.

An entrepreneurial management of a tourist enterprise can enhance its own competitiveness through specialization, innovation, investment, risk taking and productivity improvements. This strategy is also useful to improve the development of the area. The same goals are also reached by adopting ethical and cooperative business practices.

Ramayah et al. state (2011, p. 413): "a tourism network is a set of formal, co-operative relationships between appropriate organizational types and configurations, stimulating inter-organizational learning and knowledge exchange and a sense of community and collective common purpose that may result in qualitative and/or quantitative benefits of business activity, and/or community nature relative to building profitable and sustainable tourism destinations".

Tourism provides an ideal context for the emergence of networking relationships.

It is a networked sector where loose clusters of organizations within a tourist area cooperate and compete in dynamic evolution.

A core element of networking in the sector is the special sense of cohesion among the organizations. This is because they share historical, social and cultural backgrounds and they are collectively involved in order to reach the goal of sustaining the competitive advantage of the destination (Morrison et al. 2004).

In the tourism sector there are often linkages where both competitive and cooperative relationships co-exist (Wang and Krakover 2008).

A cooperative approach is required due to the characteristics of the sector and variety of stakeholders and operators that are involved. The operators need to collaborate and network in order to achieve common goals and to determine a successful sector performance and area development (Ramayah et al. 2011). The common goals may be both tangible and intangibles: for example, the former may be financial, while the latter may be knowledge sharing.

The actors involved in a tourism network develop a collective vision. To search for this common goal, they mobilize resources, develop widespread ideas and engage in cooperative actions for mutual benefit. Additionally, for a successful collaborative action the following factors are required: a high level of trust and a collective vision and commitment among stakeholders.

Komppula (2014, p. 363) defines strategic networks in the tourism sector as "strategic networks refer to shared vision and a system orientation to achieve common objectives, which requires trust and commitment among stakeholders as well as recognizing their interdependence".

The relevance of networking in the tourism sector refers also to the many SMEs in the sector which are not often growth oriented and not entrepreneur managed.

An entrepreneurial orientation allows business owners to recognise that many of the skills and resources leading to the success of their own business exist outside of the firm. Therefore, they are positively oriented to cooperation.

However, even if individual tourism businesses were not entrepreneurial, it is possible for them to join networks to allow them to promote and develop their local area. In this case a network or business association is collectively entrepreneurial.

Through networks, tourism businesses that would normally work in isolation may be involved in developing successful tourism products (Novelli et al. 2006).

The network also has the purpose of informing on the availability of certain activities in a tourist area, an activity that is even more important for new touristic destinations. Additionally, networks may increase local pride and appeal to inward investors and so positively influence the attractiveness of the area and the performance of the whole sector.

In recent decades, information and communication technology development represents the most important innovation process in the tourism sector. Information technology has been applied in the management of enterprise activity and of relations with consumers and other sector operators.

This introduced a new opportunity for collaboration among all the stakeholders that operate in the sector. Through information and communication technologies the relationships structure was redefined reducing the management costs and also increasing the value creation for the customers.

On examining the factors that have a significant influence on collaboration among tourism operators in an island economy, Ramayah et al. (2011) showed that the most important factors are communication and commitment.

Effective communication among all the stakeholders, both business owners and public institutions, may result in a better and extended collaboration among the network members.

A high degree of commitment among network members is an important factor to guarantee a successful collaborative network. Commitment, support and increased trust are important but are not in themselves sufficient as risk in the tourism network is relatively low (Medina-Muñoz and García-Falcón 2000).

Networks and collaboration are even more important for rural tourism performance (cfr. among others: Tinsley and Lynch 2001; Wang and Fesenmaier 2007).

Saxena and Ilbery (2008) point out that, in a rural context the main type of cooperation among tourism businesses was informal and based on friendship and trust. In several cases the informal cooperation represented the only form of networking. Often this is the case of emerging rural touristic areas.

Moreover, in the tourism sector, and even more so in rural tourism, also the relationships between different sectors are very significant, particularly those activated between different economic sectors, for example hospitality, agriculture and typical products.

In the past these sectors had no special reason for dialogue and convergence. Nowadays, however, they seek elements of homogeneity and dialogue-oriented joint-value creation because of an increasing demand oriented to a contextual responsible use of the land.

Hospitality, agriculture and typical products share common resources such as historical, cultural and social factors that are interwoven to create a territorial identity in the typical kinds of production and in the tourism services. The focus of their relationship, in fact, is represented by shared elements and values, such as: (1) the reference area; (2) the identity and culture of the places; (3) local sustainable development.

A collective strategy can be founded on the creation of a network among business units, whether it be a consortium, association, etc. whose goal is to achieve promotional and marketing planning, to valorise local products and tourism of a given territory.

Relations with public institutions are also relevant and their involvement focuses on the promotion and development of the area.

Management activities such as overall destination image, or international awareness of the destination are primarily the responsibility of the public sector, while the entrepreneurial quality of tourism businesses and the cooperative behaviour of the firms are the responsibility of the private sector.

From this perspective, these kinds of networks are identified as "soft" (Glosvik 2003; Rosenfeld 2001) i.e. less economic and profit-related, strongly based on social norms and reciprocity. They have open membership, involving both public and any kind of private stakeholder, and tackle generic issues.

The role of the state and local governments and the links with business units are important for sector performance because many entrepreneurs tend to start their business in their home region due to various social ties, and they are mostly oriented in a local dimension. In this case governments may contribute to supporting the private sector.

As to rural tourism, Saxena and Ilbery (2008, p. 234) have introduced the notion of integrated rural tourism. This concept identifies a tourism form which is:

> ... mainly sustained by social networks that explicitly link local actors for the purpose of jointly promoting and maintaining the economic, social, cultural, natural, and human resources of the localities in which they occur.

The authors' intention is to better underline the central role of a network connection which is useful in a rural context where sustainability is required.

The relationship would involve—with a different level of integration—all the different tourist actors as well as social, cultural, economic and environmental resources and also the end product that the amalgamation of their activities engender.

6 Conclusion

The tourism sector is changing at great speed because a greater change is influencing the national market conditions, in terms of lifestyles and consumer behaviour and international markets.

The relations between competitor countries are more and more complex, even within the EU, and this is increasing a mass demand from the new emerging countries. This situation changes the economic importance of tourism and therefore it is necessary to look at the crucial issue of resources and policies in a new way.

In Italy, tourism supply has grown in recent decades in a totally spontaneous way, more than in other economic sectors. Its growth followed the typical model of national economic development: small subjects, family-run businesses, imagination in the way of performing the service. For some time Italian tourism has been distanced from its competitors, also due to a historical, artistic and unique natural environment. More recently, when comparing Italy with its direct competitors, it emerges that our country is registering a downward trend in its share in the world tourism market. This is partly due to the enlargement of international supply but mainly to the lack of national conditions oriented to tourism. Another reason for this trend is the lack of a proper, structured tourism policy.

The economic and financial crisis that has swept across the euro-zone in recent years has led to a substantial decline in GDP and consumption. In this delicate phase, more than in others, it is important to find ways out which are new and feasible.

Although the economic and financial crisis has had negative repercussions also on the tourism sector, both on the propensity of Europeans to travel across the border and on the propensity of non-European tourists to choose Europe, this sector continues to be one of the most important on which we should focus.

As a consequence of developments in international trade and exports, in Italy a possible new way to accompany the traditional support offered to manufacturing exporters is emerging. The focus could be an action aimed at increasing the added value of the service sector.

From this point of view a simple option would be to develop tourist services, such as hotels, restaurants, bars, and all other companies located in our area, that sell to non-residents.

With reference to Italy, there appears to be a process of diversification of the tourism sector's supply with particular reference to new forms of non standardized tourism. This represents an opportunity for the whole sector to be less sensitive to the crisis.

Tourism is characterized by generating direct, indirect and induced effects in the economy both at local and national levels. Therefore, it is important for its potential role in the economic growth and development of regions within a country. Nevertheless, in Italy it appears that the role of tourism tends to be underpowered, with specific reference to the difference between the expenditure of tourists in Italy and how Italians spend abroad, as a percentage of GDP.

The characteristics of the tourism enterprises and their small scale may be a way to deal with economic drawbacks and may represent an asset in terms of implementing resilience strategies against the crisis.

With reference to the rural contexts, it appears that the global crisis may not only have relevant negative impacts on areas, but also represent important opportunities for repositioning rural areas and rural tourism enterprises and, last but not least, to contribute to driving the sector out of the crisis.

We believe that the presence of entrepreneurs among the business owners may significantly contribute to the creation of an entrepreneurial environment where the focus shifts from the growth of individual businesses to the development of the sector and of the area. Moreover, an entrepreneurial orientation is needed not only to reach the goal of growth but also to take into account the sustainable dimension of the development process from an environmental and social point of view.

As has been pointed out, lifestyle entrepreneurs, who are non-growth oriented, are present in the tourism sector. Therefore, the adoption of cooperative business practices within a network (which involves private and public stakeholders) may be useful to achieve common objectives, due to the fact that the network is collectively entrepreneurial. Moreover, in a crisis context, it represents an efficient way to purse the sector's robust and resilient performance while also taking sustainability into account.

References

Ateljevic I, Doorne S (2000) Staying with the fence: lifestyle entrepreneurship in tourism. J Sustain Tour 8(5):378–392. doi:10.1080/09669580008667374

Bruyat C, Julien PA (2000) Defining the field of research in entrepreneurship. J Bus Ventur 16:165–180. doi:10.1016/S0883-9026(99)00043-9

Busby G, Rendle S (2000) The transition from tourism on farms to farm tourism. Tour Manag 21(8):635–642. doi:10.1016/S0261-5177(00)00011-X

Cawley M, Gillmor DA (2008) Integrated rural tourism: concepts and practice. Ann Tour Res 35(2):316–337. doi:10.1016/j.annals.2007.07.011

Chuang ST (2013) Residents' attitudes toward rural tourism in Taiwan: a comparative viewpoint. Int J Tour Res 15(2):152–170. doi:10.1002/jtr.1861

Dernoi LA (1991) About rural and farm tourism. Tour Recreat Res 16(1):3–6

Dimitrovski DD, Todorović AT, Valjarević AD (2012) Rural tourism and regional development: case study of development of rural tourism in the region of Gruţa, Serbia. Proc Environ Sci 14:288–297. doi:10.1016/j.proenv.2012.03.028

European Commission (1986) Action in the field of tourism. COM (86) 32 final, Brussels, 5 Feb 1986. http://aei.pitt.edu/5411/1/5411.pdf

European Commission (2006) A renewed EU tourism policy: towards a stronger partnership for European tourism. COM (2006) 134 final, Brussels, 17 Mar 2006. http://eur-lex.europa.eu/LexUriServ/LexUriServ.do?uri=COM:2006:0134:FIN:en:PDF

European Commission (2007) Agenda for a sustainable and competitive European tourism. COM (2007) 621 final, Brussels, 19 Oct 2007. http://eur-lex.europa.eu/LexUriServ/LexUriServ.do?uri=COM:2007:0621:FIN:EN:PDF

European Commission (2010) Europe, the world's No. 1 tourist destination: a new political framework for tourism in Europe. COM (2010) 352 final, Brussels, 30 June 2010. http://eur-lex.europa.eu/LexUriServ/LexUriServ.do?uri=COM:2010:0352:FIN:en:PDF

Eurostat (2013) Eurostat news releases, no. 59/2013. http://epp.eurostat.ec.europa.eu/cache/ITY_PUBLIC/4-15042013-BP/EN/4-15042013-BP-EN.PDF

Farmaki A (2012) An exploration of tourist motivation in rural setting. The case of Troodos, Cyprus. Tour Manag Perspect 2(3):72–78. doi:10.1016/j.tmp.2012.03.007

Fayos-Solá E (1996) Tourism policy: a midsummer night's dream? Tour Manag 17(6):405–412. doi:10.1016/0261-5177(96)00061-1

Frederick M (1992) Tourism as a rural development tool: an exploration of the literature, vol 22. US Department of Agriculture, Economic Research Service, Washington, DC

Getz D, Carlson J (2000) Characteristics and goals of family and owner-operated businesses in rural tourism and hospitality sectors. Tour Manag 2(1):547–560. doi:10.1016/S0261-5177(00)00004-2

Getz D, Petersen T (2005) Growth and profit-oriented entrepreneurship among family business owners in the tourism and hospitality industry. Int J Hosp Manag 24(2):219–242. doi:10.1016/j.ijhm.2004.06.007

Glosvik Ø (2003) Networks and learning in a rural development program. Paper presented at the reinventing regions in the global economy conference, Pisa, Italy 12–15 April 2003

Johns N, Mattsson J (2005) Destination development through entrepreneurship: a comparison of two cases. Tour Manag 26(4):605–616. doi:10.1016/j.tourman.2004.02.017

Komppula R (2014) The role of individual entrepreneurs in the development of competitiveness for a rural tourism destination: a case study. Tour Manag 40:361–371. doi:10.1016/j.tourman.2013.07.007

Lane B (1994) What is rural tourism? J Sustain Tour 2(1–2):7–21. doi:10.1080/09669589409510680

Lashley C, Rowson B (2010) Lifestyle businesses: insights into Blackpool's hotel sector. Int J Hosp Manag 29(3):511–519. doi:10.1016/j.ijhm.2009.10.027

Lerner M, Haber S (2001) Performance factors of small tourism ventures: the interface of tourism, entrepreneurship and the environment. J Bus Ventur 16(1):77–100. doi:10.1016/S0883-9026(99)00038-5

McGehee NG, Kim K (2004) Motivation for agri-tourism entrepreneurship. J Travel Res 43(2):161–170. doi:10.1177/0047287504268245

Medina-Muñoz D, García-Falcón JM (2000) Successful relationships between hotels and agencies. Ann Tour Res 27(3):737–762. doi:10.1016/S0160-7383(99)00104-8

Morrison A (2006) A contextualisation of entrepreneurship. Int J Entrep Behav Res 12(4):192–209. doi:10.1108/13552550610679159

Morrison AJ, Baum T, Andrew R (2001) The lifestyle economics of small tourism businesses. J Travel Tour Res 1(1–2):16–25 ISSN 1302-8545

Morrison A, Lynch P, Johns N (2004) International tourism networks. Int J Contemp Hosp Manag 16(3):198–204. doi:10.1108/09596110410531195

Novelli M, Schmitz B, Spencer T (2006) Networks, clusters and innovation in tourism: a UK experience. Tour Manag 27:1141–1152. doi:10.1016/j.tourman.2005.11.011

Oppermann M (1996) Rural tourism in southern Germany. Ann Tour Res 23(1):86–102. doi:10.1016/0160-7383(95)00021-6

Park DB, Yoon YS (2009) Segmentation by motivation in rural tourism: a Korean case study. Tour Manag 30(1):99–108. doi:10.1016/j.tourman.2008.03.011

Peters M, Frehse J, Buhalis D (2009) The importance of lifestyle entrepreneurship: a conceptual study of the tourism industry. Pasos 7(2):393–405 ISSN 1695-7121

Phillip S, Hunter C, Blackstock K (2010) A typology for defining agritourism. Tour Manag 31(6):754–758. doi:10.1016/j.tourman.2009.08.001

Poon A (1993) Tourism, technology and competitive strategies. CAB, Wallingford

Ramayah T, Lee JWC, In JBC (2011) Network collaboration and performance in the tourism sector. Serv Bus 5(4):411–428. doi:10.1007/s11628-011-0120-z

Roberts L, Hall D (2001) Rural tourism and recreation principles to practice. CABI, Wallingford

Roberts L, Hall D (2004) Consuming the countryside: marketing for rural tourism. J Vacat Mark 10(3):253–263. doi:10.1177/135676670401000305

Rosenfeld S (Sep 2001) Networks and clusters: the Yin and Yang of rural development proceedings: rural and agricultural conferences, pp 103–120

San Martin H, Herrero A (2012) Influence of user's psychological factors on the online purchase intention in rural tourism. Tour Manag 33:341–350. doi:10.1016/j.tourman.2011.04.003

Saxena G, Ilbery B (2008) Integrated rural tourism. A border case study. Ann Tour Res 35(1):233–254. doi:10.1016/j.annals.2007.07.010

Shane S, Venkataraman S (2000) The promise of entrepreneurship as a field of research. Acad Manag Rev 25(1):217–226. doi:10.2307/259271

Sharpley R (2002) Rural tourism and the challenge of tourism diversification: the case of Cyprus. Tour Manag 23(3):233–244. doi:10.1016/S0261-5177(01)00078-4

Tajeddini K (2010) Effect of customer orientation and entrepreneurial orientation on innovativeness: evidence from the hotel industry in Switzerland. Tour Manag 31(2):221–231. doi:10.1016/j.tourman.2009.02.013

Tinsley R, Lynch PA (2001) Small tourism business networks and destination development. Int J Hosp Manag 20(4):367–378. doi:10.1016/S0278-4319(01)00024-X

UNWTO—World Tourism Organization (2012) Tourism highlights, 2012 edn. http://mkt.unwto.org/sites/all/files/docpdf/unwtohighlights12enhr.pdf

Wagener S, Gorgievski M, Rijsdijk S (2010) Businessman or host? Individual differences between entrepreneurs and small business owners in the hospitality industry. Serv Ind J 30(9):1513–1527. doi:10.1080/02642060802624324

Wang Y, Fesenmaier DR (2007) Collaborative destination marketing: a case study of Elkhart county, Indiana. Tour Manag 28(3):863–875. doi:10.1016/j.tourman.2006.02.007

Wang Y, Krakover S (2008) Destination marketing: competition, cooperation or competition? Int J Contemp Hosp Manag 20(2):126–141. doi:10.1108/09596110810852113

Williams AM, Shaw G, Greenwood J (1989) From tourist to tourism entrepreneur, from consumption to production: evidence from Cornwall, England. Environ Plann A 21(12):1639–1653. http://www.envplan.com/abstract.cgi?id=a211639

WTO—World Trade Organization (2013) Annual report 2013. http://www.wto.org/english/res_e/booksp_e/anrep_e/anrep13_e.pdf